Divining Science

Studies in Central European Histories

Edited by

Thomas A. Brady, Jr., University of California, Berkeley
Roger Chickering, Georgetown University

VOLUME LII

Divining Science

Treasure Hunting and Earth Science in Early Modern Germany

By
Warren Alexander Dym

BRILL

LEIDEN • BOSTON
2011

This book is printed on acid-free paper.

On the cover: A mine surveyor overlooks from a nearby hill, as a digger, directed by the dowser, hammers in stakes. From Balthasar Rößler's mining manual, *Bergbauspiegel* (1700). With the kind permission of the Leipzig University Library (Albertina), special collections department.

Library of Congress Cataloging-in-Publication Data

Dym, Warren Alexander.
Divining science : treasure hunting and earth science in early modern Germany / by Warren Alexander Dym.
p. cm. -- (Studies in central European histories ; v. 52)
Includes bibliographical references and index.
ISBN 978-90-04-18642-2 (hardback : alk. paper)
1. Metallurgy--Germany--Saxony--History--18th century. 2. Dowsing--Germany--Saxony--History--18th century. 3. Geochemical prospecting--Germany--Saxony--History--18th century. 4. Geophysics--Germany--Saxony--History--18th century. I. Title. II. Series.

TN642.G4D96 2010
133.3'233094321--dc22

2010028528

ISSN 1547-1217
ISBN 978 90 04 18642 2

CONTENTS

LIST OF FIGURES

ACKNOWLEDGEMENTS

This book began as a dissertation project at the University of California, Davis, expanded during post-doctoral research at Wolfenbüttel and Halle, and assumed its final form in an office at Bucknell University, where I was a visiting professor from 2007 to 2009. Numerous mentors, colleagues, and students have assisted along the way, but I can only thank a handful here.

Kathy Stuart, a true artist of the microhistorical, first sparked my passion for combing through boxes and files in the archives. It has been a pleasure to graduate from student to colleague and friend. Other important influences that date to my doctoral years include Bill Hagen, Deb Harkness, and Patrick Carroll. A special thanks to Joan Cadden for steeping me in the history of science and providing advice over the years—as her many students know, Joan also teaches the values and goals of the profession. Manfred Rudersdorf at the University of Leipzig helped integrate me into Saxon academic culture and directed my course through Saxony's libraries and archives. In 2006, I had the pleasure to meet Mary Lindemann in Wolfenbüttel, who read an early version of this work and helped steer it toward completion. I will be paying her back in Duckstein for many years to come. Ann Tlusty enriched my stay at Bucknell enormously, offering kind words of advice, becoming a fast friend and colleague. For more informal conversation and correspondence during this project I would like to thank Hjalmar Fors in Sweden, Andre Wakefield, Ernie Hamm, Bill Newman, Bruce Moran, Jole Shackelford, Mary Terral, Kathy Olesko, and Heinz Schott in Bonn. The special collections staff at the Albertina Library in Leipzig, the archivists at the Freiberg Mining Archive, and the staffs of the Government Library and Government Archive at Dresden all assisted in countless ways.

A number of institutions provided financial assistance without which none of this would be possible. Grants from the German Academic Exchange Program (DAAD) and University of Erfurt (Carl Schurz Stipend) helped jumpstart the project. I am also greatly indebted to the Herzog August Bibliothek in Wolfenbüttel and Franckesche Stiftungen at Halle for their financial help. Bucknell University provided a much-needed international travel grant during the writing phase. I should also thank the Johns Hopkins University Press for allowing a reprint of "Miners and Scholars: Dowsing and the Freiberg

Mining Academy" (*Technology and Culture* 49:4, 2008), which appears here in updated form as Chapter Six.

We often forget how our personal and emotional worlds inform our professional lives, and so I must acknowledge my parents for keeping me grounded and remaining unconditional supporters. It is to them, then, that I dedicate this book.

GLOSSARY OF MINING AND METALLURGICAL TERMS

Berg	mountain
Bergmann	miner
Bergmännlein	mining dwarf or spirit
Bergamt	local mining jurisdiction
Bergbuch	mining book
Bergfreiheit	the formal privileges of independent prospecting and digging
Bergmeister	overseer of production, manager of mines
Bergordnung	mining ordinance
Bergrat	Mining Administration at Dresden
Bergverständiger	knower of the mountain, experienced miner
Bergwissenschaft	mining knowledge
Erz	mineral ore
Erzgebirge	mountain range between Saxony and the Czech Republic (Bohemia)
Gang	mineral vein
Gebirge	mountain range
Geschiebe	alluvial mineral fragment
Gewerckschaft	collective of shareholders
Hauer	digger
Hütte	metal works
Kluft	fissure in rock or earth, often filled with mineral-bearing substance
Lachter	unit of length, ~2 meters
Lot	unit of mass, ~16 g
Marckscheider	mine surveyor
Oberbergamt	Central Mining Office at Freiberg, oversaw all jurisdictions
Oberberghauptmann	Head of the Mining Office
Probierer	assayer
Salz	salt
Sieden	boiling of brine to produce salt
Sole	brine, mineral waters
Steiger	foreman of a tunnel
Schichtmeister	shift manager
Schürf	a dig
Schürfen	digging
Schatzgraben	treasure hunting
Streich	strike; direction of the line of intersection of a vein with true horizontal plane
Wünschelrute	divining rod, dowsing rod, *virgula divina*
Witterung	mineral vapor

INTRODUCTION

> The historian is not unlike the traveler...he tends to linger over the plain, which is the setting for the leading actors of the day, and does not seem eager to approach the high mountains nearby.[1]

In a portrait hanging in the Freiberg University of Mining and Technology in Saxony, Germany, the sixteenth-century overseer of mines, Christoph von Schönberg (1554–1608), is holding a divining rod made of gilded hazel wood [fig.1]. He used the instrument to symbolize his authority over production and special access to the knowledge and experience of miners, who would have understood a natural attraction between branches and earths, and between gold and metals. In the nineteenth century, the professor of physics at Freiberg, Ferdinand Reich, examined a local mine surveyor who carried a divining rod made of iron and copper. This man claimed to detect 'galvanic excitement' above mineral ore with his metallic rod.[2] Divining or dowsing for minerals had transformed from a study of mineral vapors to the cutting-edge of electrical theory. How did pre-modern magic survive the centuries separating Schönberg and Reich to reemerge as an experimental branch of geophysics?

While mining and economic historians, legal historians, and historians of geology have focused on major changes in mining, folklorists have discovered continuity. Central European mining towns were at the crossroads of change and tradition: the expansion of state bureaucracy, wage labor and capital investment, social disciplining and reform of popular culture, and new earth science and machines; but also a proud mining culture, complete with mining songs, miner language (*Bergsprache*), processions, special garb, taboos, and orally-transmitted

---

[1] Fernand Braudel, *The Mediterranean and the Mediterranean World in the Age of Philip II*, Vol. I (Berkeley: University of California Press, 1995), 29.

[2] The Schönberg were the biggest landowners in Saxony and investors in mining. Albert Fraustadt, *Geschichte des Geschlechtes von Schönberg Meissnischen Stammes*, Band I (Leipzig, 1878). Ferdinand Reich was joint-discoverer of the element indium and one of the earliest geophysicists. Constantin Täschner, "Ferdinand Reich, 1799–1884: Ein Beitrag zur Freiberger Gelehrten- und Akademiegeschichte," *Mitteilungen des Freiberger Altertumvereins* 51 (1916). See also Chapter Six.

Fig. 1. The overseer of mines, Christoph von Schönberg (1554–1608), depicts his authority over miners and access to their knowledge by holding a divining rod.

knowledge and techniques. The diviner stood at the intersection of elite and popular cultures. All parties in mining had a stake in developing practical knowledge for the discovery of mineral resources, and men of different professional interests and social standings converged on the matter of dowsing. The practice could become new science, popular superstition, or the mark of the experienced miner, depending on who was doing the judging.

To understand how divining became experimental geophysics, we must recover diverse arguments, both for and against, but also refrain from assuming the triumph of science over superstition. Dowsing practice was surprisingly high profile, as Schönberg's portrait would suggest. The practical need to locate mineral resources and desire to develop an empirical science of prospecting benefited dowsers. Unlike alchemy and astrology, which declined in status through the eighteenth century, dowsing found an audience and patronage by mining administrations, and leading natural philosophers and scientists debated the merits of the practice. It was a part of mining science (*Bergwissenschaft*) more broadly. The mining books that presented this knowledge spoke of prospecting practices and beliefs collectively as 'digging' (*Schürfen*). Digging had a close affinity with more formal fields that emerged through the eighteenth century, such as physical geography, mineralogy, and geology (*Geognosie*), but it was a 'watered-down,' or vernacular synthesis of their most practical components. Neither did geophysics eradicate dowsing on the mountain, but provided a more powerful discourse that dowsers could co-opt. To explain the unique historical trajectory by which dowsing became an experimental or amateur science, we must remain symmetrical in our treatment of earth science and dowsing, exploring similar causal factors behind the development of each.[3]

Digging or prospecting involved how to locate, track, and extract mineral earths or springs on a small scale or exploratory basis. The experienced miner and prospector, or 'knower of the mountain' (*Bergverständiger*) studied the landscape, tasted the waters, and touched

---

[3] David Bloor, *Knowledge and Social Imagery* (Chicago: University of Chicago Press, 1991), 175–179. Bloor's symmetry postulate holds that we ought to explain both seemingly true and false beliefs with the same sorts of causes, respectively. We tend rather to explain the false or irrational with social and cultural factors, and the true or rational by assuming its inherent value, logic, or correspondence with reality. By discussing a common mining culture behind earth science and dowsing in Saxony, our study is an exercise in methodological symmetry.

stones to his tongue. He believed that fumes and miasmas, which were released in the course of mineral generation and decay, exuded upwards, providing additional signs as to where to find ore. The vegetation might appear decayed or stunted, or morning dews might evaporate more quickly than otherwise. One looked for mineral fragments (*Geschiebe*) that waters had ripped from veins and deposited at some distance from their source, and tracked them uphill, back to their origins. Since trees and bushes had great sympathy with the grounds from which they derived nourishment, the *Bergverständiger* also tested divining rods. A properly held rod might detect mineral fumes and vapors that were otherwise invisible, and lead the sensitive practitioner to suggestive spots. This theory of mineral vapors was related to contemporary alchemical and astrological conceptions of the genesis of metals. Pre-modern digging and dowsing drew on a variety of learned and popular discourses.

The patronage of *Bergverständige* by mining administrations was a function of the increasing interest European states had in craftsmen more broadly after the fifteenth century. Historians of science have shown that elites forged new associations with artisans, including miners and metallurgists. The literature has charted the improving status of *ars* among princes, humanists, natural philosophers, and state bureaucrats, and the role of craftsmen from brewers to pyrotechnicians in the development of the sciences.[4] Scholars of the Scientific Revolution

[4] Jim Bennett, "Robert Hooke as Mechanic and Natural Philosopher," *Notes and Records of the Royal Society* 35 (1980), 33–48. Alan Gabbey, "Between Ars and Philosophia Naturalis: Reflections on the Historiography of Early Modern Mechanics," in J.V. Field and Frank A.J.L James, eds., *Renaissance and Revolution: Humanists, Scholars, Craftsmen, and Natural Philosophers in Early Modern Europe* (Cambridge, MA, 1993). William Eamon, *Science and the Secrets of Nature: Books of Secrets in Medieval and Early Modern Culture* (Princeton, 1994). Simon Schaffer, "Experimenters' Techniques, Dyers' Hands, and the Electric Planetarium," *Isis* 88 (1997), 456–483. Pamela Long, "Power, Patronage, and the Authorship of Ars: From Mechanical Know-how to Mechanical Knowledge in the Last Scribal Age," *Isis* 88 (March, 1997). Pamela Smith and Paula Findlan, eds., *Merchants & Marvels: Commerce, Science, and Art in Early Modern Europe* (New York, 2002). H. Otto Sibum, "Experimentalists in the Republic of Letters," *Science in Context* 16:1/2 (2003), 89–120. Pamela Smith, *The Body of the Artisan: Art and Experience in the Scientific Revolution* (Chicago, 2006). Simon Werrett, "Explosive Affinities: Pyrotechnic Knowledge in Early Modern Europe," in *Making Knowledge in Early Modern Europe: Practices, Objects, and Texts, 1400–1800*, Pamela H. Smith and Benjamin Schmidt, eds. (Chicago: The University of Chicago Press, 2007). Some of this work builds on the classic studies of sociologists Edgar Zilsel and Boris Hessen. See Clifford D. Conner, *A People's History of Science: Miners, Midwives, and 'Low Mechanicks'* (New York: Nation Books, 2005).

defended mathematics and experimentation as appropriate for nobility, morally uplifting enterprises, even as means toward the 'good life.'[5] But the tacit knowledge of prospecting was not identical with craft experience or experimentation, though it might incorporate basic assaying and digging. The hired prospector had long experience on the mountain and an ineffable 'feel' for the terrain. He integrated treasure hunting, know-how, and experience in mining. He was part treasure hunter and part miner, and he had acquired his knowledge from miner lore, individual prospecting, as well as formal training at the mines. Dowsing and other means for locating and identifying mineral ore drew on popular, artisanal, and learned sources. For this reason we must be cautious in identifying dowsing as 'craft' or 'artisanal' knowledge per se, although we will certainly meet dowsers who had formal training in digging, assaying, surveying, and other mining-related trades.[6]

Prospecting knowledge proved quite resilient through the eighteenth century. Mining officials employed and evaluated dowsers in the field, even as bureaucrats reformed and rationalized mining, and as scholars attacked mining superstitions and imagined mining sciences. The Freiberg Mining Academy (f.1765) institutionalized a new ideal at some odds with miner beliefs, and the earliest professors attempted to systematize and theorize prospecting, among other fields. They often ridiculed miner beliefs in their scholarly texts. But a new science of

---

[5] Matthew Laurence Jones, *The Good Life in the Scientific Revolution: Descartes, Pascal, Leibniz and the Cultivation of Virtue* (Chicago: University of Chicago Press, 2006).

[6] Michael Polanyi, *Personal Knowledge: Towards a Post-Critical Philosophy* (Chicago, 1958). Tacit knowledge is pre-logical know-how that cannot be transmitted orally or in written form. It operates through idiosyncratic emotions, hunches, and guesswork. Thomas Kuhn developed the notion to help explain the transmission of scientific paradigms from teachers to students (*The Structure of Scientific Revolutions*, Chicago: The University of Chicago Press, 1996 [orig.1962], especially Chapter Five); Harry Collins employed the theory in his analysis of experimental replication failure ("The TEA Set: Tacit Knowledge and Scientific Networks," *Science Studies* 4 (1974), 165–86; and Douglas A. Harper explained the tacit knowledge of a Saab mechanic in *Working Knowledge: Skill and Community in a Small Shop* (Chicago: The University of Chicago Press, 1987). For a more recent discussion, see Harry Collins, *Gravity's Shadow: The Search for Gravitational Waves* (Chicago: The University of Chicago Press, 2004), 608–612. We refer to tacit knowledge that was not confined to the laboratory setting, mining administrations, or metallurgist's shops, but also dispersed horizontally among unlettered wanderers and farmers. See Myles Jackson, *Spectrum of Belief: Joseph von Fraunhofer and the Craft of Precision Optics* (Cambridge, MA: MIT Press, 2000), 10–13.

prospecting comparable to chemical mineralogy, geognosy, or mine engineering was not forthcoming, and instructors still used the mining books of old. Some imagined a new science of divining, as electrical currents replaced the mineral vapors of old. The pre-modern dowser spoke of mineral fumes and a God-given talent to zero-in on subterranean treasures, but his successor spoke of receptivity to galvanic or electrical impulses from below. Romantic scientists (*Naturphilosophen*) like Johann Wilhelm Ritter (1776–1810) proposed a science of dowsing in these terms, one that found an audience at Freiberg. As chemistry undermined the spiritual component of alchemy and belief in transmutation, and as astrology lost artisanal and learned foundations, dropped by elite culture, dowsing continued to find support in mining and learned circles as an experimental or amateur science.[7]

In her contribution to the *Cambridge History of Science*, Patricia Fara calls for more context-sensitive research on such forgotten Enlightenment-era beliefs as dowsing.[8] This book attends to her call. The dowsing rod was a focal point for diverse discourses—theological, scientific, artisanal, and gendered—and it serves as my point of entry, as a hermeneutic tool, into mining culture and earth science in early-modern Germany.

[7] On the decline of alchemy through the eighteenth century, see Claus Priesner, "Alchemy and Enlightenment in Germany: Ideas, Biographies, Secret Societies and a Changing Cultural Context," in Lawrence Principe, ed., *Chymists and Chymistry: Studies in the History of Alchemy and Early Modern Chemistry* (Sagamore Beach: Science History Publications, 2007). Also, Lawrence M. Principe and William R. Newman, "Some Problems with the Historiography of Alchemy," in William R. Newman and Anthony Grafton, eds, *Secrets of Nature: Astrology and Alchemy in Early Modern Europe* (Cambridge, MA: MIT Press, 2001). The alchemical work of Boyle and Newton rendered an "older," more spiritual component of alchemy less fashionable. See Betty Jo Teeter Dobbs, *The Foundations of Newton's Alchemy, or 'The Hunting of the Greene Lyon'* (New York: Cambridge University Press, 1975). On astrology in the eighteenth century, see Mark Harrison, "From Medical Astrology to Medical Astronomy: Sol-Lunar and Planetary Theories of Disease in Britain," *British Journal for the History of Science* 33:1 (2000), 25–48. Curry argues that the separation of high and low cultures gave astrology a 'plebian' status by the late eighteenth century. Patrick Curry, *Power and Prophecy: Astrology in Early Modern England* (Princeton: Princeton University Press, 1989). See also Lorraine Daston and Katherine Park, *Wonders and the Order of Nature, 1150–1750* (New York: Zone, 2001), especially Chapter Nine on the Enlightenment.

[8] Patricia Fara, "Marginalized Practices," in *The Cambridge History of Science, Vol. 4, Eighteenth-Century Science*, ed. Roy Porter (Cambridge: Cambridge University Press, 2003), 491.

Mining and Modernization

Mining in the Ore Mountains (*Erzgebirge*) has been an important subject in German economic history since the nineteenth century. The literature positions early mining centrally into arguments on modernization and industrialization, both Marxist and neo-classical, which a study of dowsing will complicate.[9] Friedrich Engels once described gold and silver production in the Erzgebirge between 1470 and 1530 as the "final moment" of the first bourgeois struggle against feudalism (the "so-called Reformation"), in which Germany saw the formation of an early proletariat, an idea that Karl Kautsky developed further in an 1889 paper.[10] East German economic historians such as Johann Köhler and Adolph Laube, much in line with Marx and Engels, emphasized the importance of accumulated capital in mining production and the development of an early proletariat in mining.[11] Eberhard Wächtler saw the birth of capitalism from feudalism in the distinctive freedoms of miners:

> So began…the history of the freedom of trade. This was not yet a function of bourgeois demands, but rather the result of the social effect of a 'modern' qualification of feudal agricultural policy and…the overcoming of feudal privileges. With this Saxon mining was marked by elements that favored bourgeois development in the city and country, and that finally proved to be the seeds of capitalism.[12]

Western scholars have revised the Marxist approach by explaining various interests in the town and countryside beyond the peasant struggle against their lords. Peter Blickle discussed the role of miners in the 1525 peasant uprisings and war. He suggested that townsmen and miners had interests similar to burghers and peasants, and preferred the title 'Revolution of the Common Man' to the more restrictive 'Peasant's War.' Miners, the "first 'modern' laborers in a predominantly

[9] The reader most interested in dowsing practice itself might skip the following literature review and proceed to the chapter overview that ends the introduction.

[10] 'Brief an Kautsky v. 15.9.1889,' in Melst, *Zur deutschen Geschichte* I (Berlin, 1957), 617. Karl Kautsky, "Die Bergarbeiter und der Bauernkrieg," *Die neue Zeit* 7 (1889).

[11] Johann Köhler, "Die Keime des Kapitalismus im sächsischen Bergbau (1168 bis um 1500)," *Freiberger Forschungshefte* D13 (Berlin: Akademie Verlag, 1955). Adolph Laube, *Studien über den erzgebirgischen Silberbergbau von 1470 bis 1546* (Berlin: Akademie Verlag, 1974).

[12] Eberhard Wächtler, "Bergbaureviere als Vorreiter technischer Entwicklungen des 18. und 19. Jahrhunderts unter besonderer Berücksightigung Sachsens," in Ekkehard Westermann, ed., *Vom Bergbau zum Industrierevier* (Stuttgart, 1995), 368.

agrarian society of estates," identified with these other groups in opposition to the emerging state.[13] The participation of miners in Saxon uprisings was sporadic, but that in the Tyrol in Austria was more organized, where several thousand miners rose up in January of 1525. "Early capitalist production had one of its centers in the gold, silver, and copper mines of the Austrian mountains, and the conflict thus brought the first 'laborers' into the movement."[14] In May 1525, rebels looted the property of the famous Fugger family, major investors and owners of smelting works. When peasants appeared for the first time at imperial diets alongside nobility, clergy, and townsmen, workers from the mining town of Schwaz joined peasants at the Innsbruck diet in June of 1525.

Yet, Hermann Löscher argued for the social distinctiveness of the guild-like miner fraternities: "When we consider this [social] compound of the Brotherhood [*Knappschaft*], it is impossible to describe it as an organization of class, much less of class struggle."[15] The Reformation had a major social impact on miners and secularized the mining confraternities, making them an arm of the state, though Karant-Nunn similarly denies that Saxon miners achieved collective class-consciousness in the sixteenth century. While she concedes some similarity between Saxon miners' labor and the later proletarian mode of production, she argues, "A proletarian mode of work tended to precede the evolution of a shared self-image among the miners."[16] Studying the English case, by contrast, Andy Wood argues that free miners had long maintained a "peculiar autonomy and sense of collectivity" since late medieval times.[17] They developed a legalistic sense and pitted customary law against legal reforms when confronted by the property interests and interventions of their rulers.

---

[13] Peter Blickle, *From the Communal Reformation to the Revolution of the Common Man* (Boston: Brill, 1998), 68.

[14] Blickle, *From the Communcal Reformation*, 112.

[15] Hermann Löscher, "Die erzgebirgischen Knappschaft vor und nach der Reformation," *Blätter für deutsche Landesgeschichte* 92 (1956), 181.

[16] Susan C. Karant-Nunn, "From Adventurers to Drones: The Saxon Silver Miners as an Early Proletariat," in *The Workplace before the Factory: Artisans and Proletarians, 1500–1800*, ed. T.M. Safley and L.N. Rosenband (Ithaca: Cornell University Press, 1993), 74. Also Karant-Nunn, "Between Two Worlds: The Social Position of the Silver Miners of the Erzgebirge, c. 1460–1575," *Social History* 14:3 (1989), and George H. Waring, "The Silver Miners of the Erzgebirge and the Peasant's War of 1525 in the light of Recent Research," *The Sixteenth Century Journal* 18 (1987).

[17] Andy Wood, "Custom, Identity and Resistance: English Free Miners and Their Law c.1550–1800," in Paul Griffiths, Adam Fox, and Steve Hindle, eds., *The Experience of Authority in Early Modern England* (New York: St. Martin's Press, 1996), 253.

Economic historians of mining are no longer so preoccupied with class-consciousness and the shift from feudal to capitalist organization, though early mining remains mostly a case study in the transition to modernity. They chart a marked upswing in central-European mining and metallurgy between the mid-fifteenth and mid-sixteenth centuries, and the progress of technology. Within this long-term cycle, Ian Blanchard specifies four medium-term cycles to account for the alternating dominance of Saxony, Thuringia, and the Tyrol.[18] A number of interrelated factors contributed to the development of this region, including expansion in trade and higher demand for coins, greater state and private investment, discovery of new veins, and new technologies to reach and work deeper deposits. The flourishing of technical literature at the turn of the sixteenth century demonstrates that the water wheel, pump, screw-press, and roller all found wider application, especially in the metals industry.[19] To extract mineral ore at depths deeper than one hundred meters, miners constructed new water wheels, supply ponds, and conduits to replace human and animal power, as illustrated most famously in Agricola's *De re metallica* (1556). The recipricating motion from water wheels could be transmitted along linkages to pumps located many meters away.[20] Steam technology replaced this 'rod engine' (*Stangenkunst*) design only in the nineteenth century. Related technological improvements in metallurgy included the use of coal as a replacement for wood, and new techniques for extracting gold from rock (mercury amalgamation) and silver from copper ore (liquation).

Mining and economic historians of eighteenth-century mining similarly focus on progress and industrialization. Freiberg innovations in particular were a watershed in mine engineering history. Johann Friedrich Mende, "founder of a new age" of machines at Freiberg, was a graduate of the Academy.[21] Freiberg ingenuity contributed to

---

[18] Ian Blanchard, *International Lead Production and Trade in the 'Age of the Saigerprozess' 1460–1560* (Stuttgart: Franz Steiner Verlag, 1995).

[19] Hermann Kellenbenz, "Technology in the Age of the Scientific Revolution," in Carlo Cipolla, ed., *The Fontana Economic History of Europe*, vol 2 (London: Fontana, 1973).

[20] Terry S. Reynolds, *Stronger Than a Hundred Men: A History of the Vertical Water Wheel* (Baltimore: The Johns Hopkins University Press, 1988), 133–136, 140–143.

[21] Hans Baumgärtel, *Bergbau und Absolutismus: Der sächsische Bergbau in der zweiten Hälfte des 18. Jahrhunderts und Maßnahmen zu seiner Verbesserung nach dem Siebenjährigen Kriege* (Leipzig, 1963), 54. Mende (d.1798) became a Director of Engineering (*Maschinendirektor*) and invented extraction machines and water pumps.

the application in mining of Thomas Newcomen's improved steam-powered engine. Another Academy graduate, Christoph Friedrich Brendel, studied this technology in England and introduced it to Freiberg in 1808. The mining administration spearheaded construction of a major steam-powered machine in 1817, and Brendel oversaw the construction of underground railroads for the hauling of ore in 1829, perhaps the "first railroad in Germany."[22] James Watt and George Stephenson sent their sons to study at Freiberg, where Abraham Werner lectured on coal mining.[23]

Early mining was certainly an arena replete with innovation, and a magnet for capital. But an exclusive focus on progressive technologies and economic cycles can suggest that miner traditions had only negative impact, or that they were necessarily at odds with capitalism, new science, and improved technologies. But might progress in Saxon mining not have demanded a renewed interest in the knowledge of miners? Miners could offer material-oriented magic like dowsing that landowners and administrators like Schönberg desired to control. The assumption of the present work is that progress in Saxon mining brought high and low cultures together around the issue of prospecting. To appreciate this convergence, we need a subtler and more context-sensitive approach to the decline of magic.[24]

The Decline of Magic

Folklorists of mining have long studied the beliefs and rituals of miners. Mining culture grew "hand in hand" with economic development

---

[22] Hanns-Heinz Kasper and Eberhard Wächtler, eds., *Geschichte der Bergstadt Freiberg* (Weimar, 1986), 186.

[23] Wächtler, "Bergbaureviere." Werner was professor at the Freiberg Academy from 1775 to 1817. He advanced the study of rock stratifications or 'formations' (*Gebirgsformation*), which he made central to geology, or what he termed '*Geognosie*.' See Chapter Six.

[24] A quick note on terminology: whenever possible, I will refer to the office or position under discussion, as in 'digger' (*Hauer*) or 'mine manager' (*Bergmeister*), mentioning relative social standing or income as appropriate. The more generic term, 'miner' (*Bergmann*), while it had great currency in the pre-modern period, obscures important distinctions. When I use the term, I am generally referring to workers required to be physically present at the mines, such as windlass-men or shift managers, rather than more office-bound, administrative, or city-based individuals. German speakers made a similar distinction between 'miners of the leather' (*Leder*) and 'miners of the quill' (*Feder*). See E.P. Hamm, "Unpacking Goethe's collections: the public

in Saxony, according to the folklorist of central European mining, Gerhardt Heilfurth.[25] Processions, festivals, iconography, songs, prayers, and votive pictures to protective saints, often displayed at the opening of shafts, all developed since medieval times. Diggers had unique clothing and carried particular tools that came to symbolize their work and ethic, especially the pick and axe. They frequently believed in mining spirits, mineral vapors, the organic growth of mineral veins, the influence of the stars, and the efficacy of the dowsing rod, among other tools of legitimate and illegitimate magic. Addressing the gradual *Christianization* of mining through medieval times, Georg Schreiber found that the naming of a mine by the major shareholders was likened to a baptism.[26]

Folklorists generally do not intend to explain change over time, and they frequently assume that modernizing influences, such as literacy and science, undermined folk beliefs. But historians of popular culture and magic have portrayed folk belief to be fundamentally opposed to science. Keith Thomas, Peter Burke, Richard van Dülmen, and others have claimed that belief in prophecy, spirits, and magic declined in an essential sense at all social levels during the Enlightenment.[27] This was due to any number of overlapping and interweaving factors, including the separation of elite and popular cultures, the social disciplining of popular culture, growing commercialization, agricultural and technological improvements, demographic changes, popular education and higher literacy rates, an experimental approach to natural phenomena, the mechanical philosophy, statistics and probability, sociology, psychology, and any combination thereof. Van Dülmen specifies the eighteenth century as an important breaking point: "Only in the eighteenth

---

and the private in natural-historical collecting," *British Journal for the History of Science* 34 (2001), 286.

[25] Gerhardt Heilfurth, *Das Erzgebirgische Bergmannslied: ein Aufriss seiner literarischen Geschichte* (Frankfurt am Main: Weidlich Reprints, 1982), p. 8.

[26] Georg Schreiber, *Der Bergbau in Geschichte, Ethos und Sakralkultur* (Cologne: Westdeutscher Verlag, 1962). On confessionalization and popular magic in Saxon mining, see Warren Alexander Dym, "Mineral Fumes and Mining Spirits: Popular Culture in the *Sarepta* of Johann Mathesius, 1504–1565," *Renaissance and Reformation Review* 8.2 (2006).

[27] Keith Thomas, *Religion and the Decline of Magic: Studies in Popular beliefs in Sixteenth and Seventeenth Century England* (New York: Oxford University Press, 1971). Peter Burke, *Popular Culture in Early Modern Europe* (London: Temple Smith, 1978). Richard von Dülmen, *Kultur und Alltag in der Frühen Neutzeit: Religion, Magie, Aufklärung* (München: C.H. Beck, 1999).

century did the rational-thinking mind free itself from the magical worldview."[28]

Historians of geology have shown that mining was as an important motivation behind geological knowledge, but they complement the folklorists and historians of popular culture by relegating miner beliefs to the pre-modern. A number of important works describe the suppression of pre-modern theories of the earth by more mechanically oriented approaches.[29] The following passages might serve as representative. David Oldroyd is highly sensitive to the contribution that miners and mining books have made in earth science and mining history, but he draws a clear distinction between the "tacit knowledge" of miners and the "philosophical mining" of the late eighteenth century:

> The miners themselves had their own tacit knowledge and their own vocabulary, some of it derived from the old organic traditions about the earth… But while miners knew how to excavate pits and wells, drive shafts, pump out mine waters and recognize signs…that might indicate profitable or unprofitable digging…mining did not, in itself, furnish new ways of thinking about the earth. The miners dug in the earth and had rules of thumb as to how to proceed… Eventually, in Germany, there arose also a new kind of 'philosophical mining,' which gave rise to the first establishment of the basic outline of the stratigraphical column by Abraham Gottlob Werner, lecturing at the Mining Academy in Freiberg.[30]

A second example comes from Martin Rudwick: "A Saxon mining foreman, and even an ordinary miner, might possess a wealth of practical knowledge about the mineral veins in the Erzgebirge, or at least those in their own mines, but could hardly form any scientific

[28] "Erst im 18. Jahrhunder, 'befreite' sich eine rational denkende Intelligenz vom magischen Weltbild." Van Dülmen, *Kultur und Alltag*, 82.

[29] David Oldroyd, *Thinking about the Earth: A History of Ideas about Geology* (Cambridge, MA: Harvard University Press, 1996). Christoph Bartels, Reinhard Feldmann, and Klemens Oekentorp, *Geologie und Bergbau im rheinisch-westfälischen Raum* (Münster: Universitäts- und Landesbibliothek Münster, 1994). Rachel Lauden, *From Mineralogy to Geology: The Foundations of a Science 1650–1830* (Chicago: The University of Chicago Press, 1987). Martin Guntau, *Die Genesis der Geologie als Wissenschaft: Studie zu den kognitiven Prozessen und gesellschaftlichen Bedingungen bei der Herausbildung der Geologie als naturwissenschaftliche Disziplin an der Wende vom 18. zum 19. Jahrhundert* (Berlin: Akademie Verlag, 1984). Eberhard Wächtler, ed., *Geologie und Industrielle Revolution: Studienmaterial zur Wissenschaftsgeschichte und Geschichte der Produktivkräfte* 2 (Freiberg: Bergakademie Freiberg, 1980).

[30] Oldroyd, *Thinking about the Earth*, 70.

generalization from it."[31] Oldroyd and Rudwick are chiefly concerned with the triumph of progressive knowledge on the mountain from the top downward, whereas we are presenting a vernacular science of mining, and arguing that new science integrated with tradition in the case of divining.

Few historians have explored the shared ground that unites folklorists and historians of geology, or how a common mining culture informed the full spectrum of offices, from diggers, to shaft managers and foremen, to mine masters and surveyors, to head overseers, to mining bureaucrats in Dresden, and even the Elector himself. It is this common ground that the present work charts, as it explains how the dowsing rod found widespread employment despite being constantly under attack, how it appropriated powerful discourses in mining and natural philosophy, and almost fulfilled the Enlightenment dream of a scientifically grounded prospecting practice.

There is no overarching explanation for the decline of popular mining culture. Different elements underwent distinct changes—belief in the generation of ore, mineral vapors, mining spirits, civic rituals and processions, miner language (*Bergsprache*), mining honor—and the trajectory of decline was not simply linear. There were retrogressions, sudden peaks in popularity, waves of intellectual interest, emerging and submerging theories, as well as full-scale attacks. Dowsing was not merely an instance of the dichotomy between science, confessionalization, or social disciplining and popular magic. It did not parallel the historical trajectories of astrology, alchemy, witchcraft, or popular beliefs like the flight of the *benandanti* of Friuli, or the phantoms of the night at Oberstdorf, which disappeared from the historical record just as dowsing became more firmly entrenched.[32] The learned community at Freiberg was hostile to many miner beliefs, including the influence of the stars in the generation of metals and mining spirits, but its position on the dowsing rod was always ambiguous.

Are there other clues in the literature as to why dowsing might have remained important through the Enlightenment? Wolfgang Behringer

---

[31] Martin J.S. Rudwick, *Bursting the Limits of Time: The Reconstruction of Geohistory in the Age of Revolution* (Chicago: The University of Chicago Press, 2005), 33.

[32] Carlo Ginzburg, *The Night Battles: Witchcraft and Agrarian Cults in the Sixteenth and Seventeenth Centuries*, John and Anne Tadeschi, trs. (Baltimore: Johns Hopkins University Press, 1992). Wolfgang Behringer, *Shaman of Oberstdorf: Chonrad Stoeckhlin and the Phantoms of the Night*, H.C. Erik Midelfort, tr. (Charlottesville: University Press of Virginia, 1998).

observes an increase in treasure hunting-related criminal investigations in eighteenth-century Bavaria. This had become a "fashionable crime" (*Modeverbrechen*) among men.[33] Eva Labouvie sees a correlation between the growing interest of the rural population in personal property and material well being, and an increasing appeal to magical treasure hunting and other individualized forms of magic concerned with material gain. By contrast, more community-oriented forms of magic geared toward other aims, such as casting spells on neighbors or healing the sick, declined.[34] The dowsing rod was a material-oriented form of magic, and it empowered the individual. The culture of mining positively valued the improvement of material wellbeing, including the purchasing of shares (*Kuxe*), and it promoted a sturdy individualistic and masculine ethic. Johannes Dillinger develops another association between treasure hunting and personal property. He sees interest in these practices to be specific to a transitional period between agrarian and market economies. Drawing on the anthropologist, George M. Foster, Dillinger argues that European treasure hunting became important in the eighteenth century because it was an acceptable means within agrarian, 'limited good' society to participate in market-oriented activity.[35] The limited good theory has appeal here, since mining was an arena in which the transition from agrarian to market economies was striking, and early prospectors were frequently also farmers.

Brotherhood and Exclusion of Women

The early sixteenth-century boom in the Ore Mountains brought new civic structures and values where once laid only farmland, thick woods,

[33] Wolfgang Behringer, *Hexenverfolgung im Bayern: Volksmagie, Glaubenseifer und Staatsräson in der Frühen Neuzeit* (Munchen: R. Oldenbourg, 1988), 348.

[34] Eva Labouvie, *Verbotene Künste: Volksmagie und ländlicher Aberglaube in den Dorfgemeinden des Saarraumes, 16.–19. Jahrhundert* (Köln: Röhring Verlag, 1992), 300.

[35] A limited good society would be a closed system, in which any individual's gain is the loss of another. Johannes Dillinger, "Treasure-Hunting: A Magical Motif in Law, Folklore, and Mentality, Württemberg, 1606–1770," *German History* 20:2 (2002), 178–184. On Foster's concept of the limited good, see George M. Foster, "Peasant Society and the Image of Limited Good," *American Anthropologist* 67 (1965), 293–315. The concept has also been applied for early American treasure seeking. See Alan Taylor, "The Early Republic's Supernatural Economy: Treasure Seeking in the American Northeast, 1780–1830," *American Quarterly* 28 (1986), 18.

and a passage to Bohemia. Peasant farmers supplemented their income with work at the mines or by purchasing shares. As miners and officials worked to formulate mining science, as the state took increasing control over production, and as workers organized into guild-like brotherhoods (*Knappschaften*) to protect their interests, women inevitably found fewer jobs at the mines. The increasingly male face of mining, and masculine honor and ethic this encouraged fostered the image of the (male) *Bergverständiger*, who demonstrated a mastery of the hidden knowledge of nature, especially when prospecting, as evidenced again by the Schönberg portrait.

Given the great social and economic fluctuations of sixteenth-century mining towns, and the high turnover rate of the work, the need to form all-male associations on the mountain was particularly acute.[36] Like the craft confraternities that preceded city guilds, the brotherhood was originally a religious organization charged with levying funds for processional candles, with organizing vigils, masses, and other ceremonies, and with caring for the sick and injured.[37] If the guild protected artisans from the practices of merchants, the miners' brotherhood was a similar response to bureaucratic expansion and the influence of outside capital in mining. Many customs of the brotherhoods were drawn from those of the guild, including the yearly communal meal, burial ceremonies, and placement in processions. Unlike a guild, however, the brotherhood began as an association of common diggers (*Häuergesellschaft*) rather than skilled artisans, and it came later to accommodate various higher-status individuals, including mine managers and investors. The brotherhood also did not institute a rigid dichotomy between masters and apprentices. The relative openness of the brotherhood in comparison to guilds encouraged the migration of miners to the mountains, whereas a guild structure would have restricted such movement, which was opposed to royal interest

[36] "Confraternities and their later incarnations provided the sense of belonging and the means for distinction in a world of fleeting encounters and ephemeral personal contacts. It is surely no accident that these brotherhoods were most prevalent among workers of stone, wood, leather, metal, and cloth, that is in trades where the pool of labor was large, the turnover high, and thus the demands for distinction most pressing." James R. Farr, *Artisans in Europe, 1300–1914* (Cambridge: Cambridge University Press, 2000), 251.

[37] Farr, *Artisans in Europe*, 230.

and policy.[38] The early brotherhoods also did not struggle with city patricians, as guilds always had.[39]

Women were involved in surface-level, family oriented mining in Saxony until around the 1580s. In the late sixteenth century, when a bushel of corn cost between 35 to 50 groschen, a common digger's weekly wage was some 21 to 23 groschen.[40] The separators and washers in the nearby huts earned a comparable wage, and the windlass men above ground earned between 12 and 15.[41] A small percentage of women at Freiberg between roughly 1517 and 1540 performed the task of mixing coal dust with clay and forming bricks from the compound for use in separating silver from copper (liquation). For such skilled labor, they earned up to 10 to 18 groschen weekly, comparable to the windlass men. More commonly, women's work was part-time and lower paying than this. The same task of mixing coal dust with clay at Schneeberg earned one woman just 2.5 groschen. More typical work included breaking, rinsing, and transporting ore, carrying equipment such as tallow, ash, and rope to the miners, and other piecemeal tasks. Women earned some pfennig or groschen for such work, the need for which varied from mine to mine, and from day to day. These tasks and their pay were not included in the mining ordinances that otherwise described all male jobs, with the exception of a

38 Hermann Löscher, "Die erzgebirgischen Knappschaften vor und nach der Reformation," *Blätter für deutsche Landesgeschichte* 92 (1956), 162–190. It is tempting to conclude that the relative openness of the *Knappschaft* vis-à-vis the city guild contributed to the higher level of technological innovation in mining, but such an argument would have to contend with the counter-charge that the guild structure fostered certain innovations. See Stephen R. Epstein, "Craft Guilds, Apprenticeship, and Technological Change in Preindustrial Europe," *The Journal of Economic History* 58:3 (Sept. 1998), 684–713.

39 Tensions did develop between incorporated and rural miners, however, and could erupt in real violence, as during the Peasant's War at Annaberg, when town-based miners demanded preference at the mines over countryside workers. Karant-Nunn adds that town-folk considered rural miners, which would have included farmers and more transitory individuals, 'dishonourable' (Karant-Nunn, "Between Two Worlds," 315, 319). Other problems could develop between peasants and miners: a landowner did not own the earth below his crops, and the independent prospector might trespass fences and walls and dig on a simple whim, in the name of the Elector; he was only required to fill his hole before leaving.

40 Kasper and Wächtler, *Geschichte der Bergstadt Freiberg*, 122. In the late sixteenth century, one Meißener Gulden=21 Groschen=252 Pfennige. Walter Schellhas, "Das sächsische Geld des 16. Jahrhunderts," *Mitteldeutsche Monatshefte* 10 (1926).

41 Kasper and Wächtler, *Geschichte der Bergstadt Freiberg*, 124.

Bohemian piece that specified that women be paid as much as the lowest-paid boy.[42]

There were scattered depictions of women performing these tasks in collections of religious songs, mining legislation, paintings, and woodcuts.[43] Agricola included 21 women in the 289 woodcuts that adorned his famous *De re metallica* (1556), among them, one woman washing ore. The other women in Agricola's images were shown engaged in various activities, some of which had no immediate importance to mining—embracing a man in secret, carrying an infant—and others of which indicated their working roles, such as eating a meal before a salt boiling shack, or sorting ore at a table. "The work of sorting the crude metal or the best ore is done not only by men, but also by boys and women."[44] With these words, Agricola may have expressed his pride that mining was so varied in its work to include individuals otherwise excluded from the labor market. Intended for wealthy investors interested in the sort of capital-intensive machinery and metallurgical processes that in turn barred women, the text would contribute to their exclusion.

Demographic expansion and inflation, and a transition to deeper-level, capital-intensive mining increasingly privileged the passing of skills through the male-dominated brotherhoods. Vanja argues that early industrialization and state capitalism excluded both women and unskilled male laborers from mining.[45] Only the Freiberg fraternity, at the center of all Saxon operations, included women just before the Reformation. The 1521 register still listed 86 'old sisters' of 436 total members. In Karant-Nunn's opinion, these were mostly the wives of respected miners, identified by their husband's names, and they would have participated in the social, spiritual, and welfare functions of the brotherhood, rather than served at the mines or been given leading

[42] Susan C. Karant-Nunn, "The Women of the Saxon Silver Mines," in *Private Lives: Women in Reformation and Counter-Reformation Europe* (Bloomington: Indiana University Press, 1989), 36–9.

[43] Christina Vanja, "Mining Women in Early Modern European Society," in Safley and Rosenband, *The Workplace before the Factory*.

[44] In Karant-Nunn, "Women of the Saxon Silver Mines," 30.

[45] Christina Vanja, "Mining Women." See also her "Bergarbeiterinnen: Zur Geschichte der Frauenarbeit im Bergbau, Hütten- und Salinenwesen seit dem späten Mittelalter, Part II," *Der Anschnitt: Zeitschrift für Kunst und Kultur im Bergbau* 40 (1988), 128–43.

roles in processions, like the men. At Annaberg, some one thousand women performed social and welfare roles for the Society of Saint Anne (patron saint of mining). The year that Duke Heinrich introduced Reformation initiatives in Saxony (1537), however, a line was drawn through all but one female member of the Freiberg brotherhood. Future mention of female members invariably referred to noblewomen, suggesting that the post-Reformation brotherhood became a more elite organization.[46]

These trends continued into the seventeenth century. "Women could obtain nothing more than…temporary work around the mines, their places having been taken by men and boys."[47] At Annaberg, one especially enterprising woman, the widow of a well-to-do burger and investor, supposedly taught a number of women how to make bobbin lace for the market, and soon up to nine hundred were involved in this task. The industry continued into the nineteenth century. This development, irrespective of its origin, is testament to the readiness of mining-town women to supplement their husbands' incomes, but also to the decreasing output of the mines after 1600 and fewer opportunities they afforded women.

The spread of the brotherhood and its male honor code paralleled other city-based trades from brewing to tanning, in which the guild structure excluded women on the pretext of skill and honor.[48] As even journeymen's position grew more precarious, they organized as *Gesellen* and further distinguished their work from that of women.[49] Some women managed to form guilds in the textiles.[50] We should remember these trends when encountering examples of miner pride and honor, the image of the *Bergverständiger*, and arguments against the claim that dowsing was a species of witchcraft.[51]

---

[46] Membership for 1542 included Elector Johann Friedrich, Duke Moritz, and the future Elector August. Karant-Nunn, "Women of the Saxon Silver Mines," 40.

[47] Karant-Nunn, "Women of the Saxon Silver Mines," 37.

[48] Merry Wiesner-Hanks, "'A Learned task and given to men alone': the gendering of tasks in early modern German cities, *Journal of Medieval and Renaissance Studies* 25:1 (Winter, 1995).

[49] Merry Wiesner, "Guilds, Male Bonding and Women's Work in Early Modern Germany," *Gender and History* 1:2 (Summer, 1989).

[50] Farr, *Artisans in Europe*, 41.

[51] While he demonstrated expertise and tacit knowledge, I find little evidence that the pre-modern *Bergverständiger* 'raped' nature, in the sense Carolyn Merchant

Chapter Overview

Chapter One presents a short case study of dowsing for salt as a means of illustrating how the dowser worked with the mine surveyor. We recreate an expedition for brine between 1713 and 1714 conducted in the Saxon countryside and authorized by Dresden. Unencumbered by any social or epistemological division between surveying and dowsing—between applied mathematics and miner lore and experience—the mine surveyor made decisions in consultation with diviners. That interaction helped distinguish dowsers from diggers, and the dowsers were paid accordingly. Chapter Two then provides a more formal introduction to this curious practice. Divining was a component of popular treasure hunting, where the rod might appear as an instrument of sorcery and witchcraft. But as witch hunting flared in Saxony, while a discourse of witchcraft emerged for dowsing, male treasure hunters and dowsers did not number among the accused. In the context of mining, the major theory that legitimized dowsing, and disassociated it from illegitimate magic was the notion that mineral ore exuded vapors that the *Bergverständiger* detected with a rod. Influential authors such as Georg Agricola and Jean Bodin helped distinguish dowsing from witchcraft, propagate mineral vapor theory, and develop the image of the *Bergverständiger*.

Chapter Three explains how mineral vapor theory became a component of mining science (*Bergwissenschaft*) after 1650. Following the Thirty Years War, bureaucrats and mining officials initiated great top-down administrative reforms, and spoke of mining science. Maintaining multiple commitments to scholars, miners, and Dresden officials, however, the reformers did not impose a mechanical philosophy or other learned theory in mining. They negotiated diverse interests to compose a vernacular science of disparate intellectual origins. Presented in their mining books, *Bergwissenschaft* incorporated the latest in the (al)chemical analysis of minerals and mathematics of mine surveying, it explained new technologies like the mining compass, and it drew on the tacit knowledge and lore of miners. This 'new science' legitimized the dowsing rod as a study of mineral vapors, and

advances in *The Death of Nature: Women, Ecology, and the Scientific Revolution* (Harper Collins, 1980) and her other works. Unlike his industrial-minded successors, his worldview was decidedly non-mechanistic, and his digging generally surface-level.

authors explained and depicted how dowsing complemented the mine surveyor's work in particular.

Chapter Four continues and expands the microhistory of dowsing practice by analyzing a series of practitioners in the decades just prior to the Freiberg Mining Academy. The correspondence between Dresden bureaucrats and Freiberg officials, formal contracts, and mining reports reveal how officials of the 1730s and 40s were becoming more skeptical of the would-be dowser than officials were in the 1713 case, but remained open-minded, as the mines became a testing-ground for the practice, and dowsers worked formally and publicly for mining administrations. We also pause here to present a profile of the formally employed dowser.

Chapter Five broadens the focus and anticipates our analysis of the Academy by asking if scholars of the Scientific Revolution generated a new mining science to challenge contemporary *Bergwissenschaft*. Natural philosophers across Europe gathered information about miners and miner beliefs, seeking reliable mineralogical knowledge. Dowsing became something of a litmus test among the learned for prospecting tradition. French Cartesians advanced a corpuscular (atomic) theory to explain one peasant practitioner's presumed success, whereas an anti-Cartesian (vitalist) alternative quickly followed in Halle, Germany. Meanwhile, Leibniz advanced the argument that new earth science should replace superstitious beliefs. His ideal of scientific knowledge and rationalized control over mining was inevitably hostile to prospecting practices such as dowsing. Mining bureaucrats and professors at Freiberg inherited this rich intellectual legacy.

Chapter Six examines the Freiberg Academy and electrical science of dowsing that came to replace mineral vapor theory. Notwithstanding the rhetoric of mining bureaucrats like Anton von Heynitz and professors like Johann Friedrich Wilhelm von Charpentier, who called for a break with miner traditions, prospecting was steeped in older *Bergwissenschaft*, and the mining books that preserved it held authority at Freiberg. German physicists and Romantic scientists, foremost among them, Johann Ritter, developed a new approach to divining that drew on Galvani's and Volta's research. The famous Freiberg physicist, Ferdinand Reich, examined a diviner who claimed to experience a galvanic response to mineral ore. This case demonstrates the new synthesis of prospecting knowledge and electrical theory that would mark divining practice in the modern age. Dowsing for mineral ore had survived the eighteenth century to appropriate the language of electrical currents and galvanism. This made an experimental science out of an age-old magic wand.

The inquisitive reader, perhaps aware of modern-day water witching and well seekers, may expect the 'truth' about dowsing in these chapters. Is the electrical theory a scientifically valid proposition? Or is dowsing rather a centuries-old fraud and self-delusion? But the reader will be disappointed in this regard, and might rather consult the research of physicists and psychologists who claim to have solved the mystery.[52] My own approach is more humble and echoes David Bloor's example of the anthropologist who studies witchcraft: "When anthropologists study...a witchcraft culture they are implicitly asking what circumstances would permit a rational person to embrace such beliefs. This question can be addressed and answered without becoming a believer. It is consistent with a residual evaluation that such beliefs are false."[53] Most dowsers and their supporters were rational individuals, and explaining why they believed the rod to dip at one spot and not at another does not require that it dip for us. The important point for the cultural historian is to explain why the actors believed what they did, not to evaluate their beliefs: "That the world does not contain witches leaves open the question of whether it will or not be believed to contain witches."[54] I confess a great amount of skepticism on the matter of dowsing, but an equally profound one for many contemporary scientific and medical practices. Meanwhile, the world continues to produce dowsers.

---

[52] William Barrett and Theodore Besterman, *The Divining Rod: An Experimental and Psychological Investigation* (London: University Books, 1926).

[53] Bloor, *Knowledge and Social Imagery*, 76.

[54] Bloor, *Knowledge and Social Imagery*, 177.

CHAPTER ONE

WHITE GOLD ON SPITZBERG HILL

This is a tale of salt. The people of the Dübener marshlands between Torgau and Wittenberg had always known that mineral springs lay untapped below. Stories had circulated among the peasants and townspeople as fluidly as did the brine itself, until one day, in the summer of 1713, rumors of salt near the town of Gommlo reached the Elector in Dresden. Expeditions were organized, workers were hired, and reports were penned of the tedious work of discovering and analyzing the mineral content of subterranean brine. In the end, very little would come of this expedition to Gommlo. Officials filed the papers away, as more productive centers opened at Artern, Kösen, and Dürrenberg after 1725. Were it not for the large surveyors' maps (*Risse*) tucked into the files, the events at Gommlo may have escaped the historian's notice as well. But the drawings and their long-winded titles can attract the attention of one seeking to understand how mineral resources were tapped, and what sort of knowledge worked in practice [Fig. 2]: "By most gracious order from Dresden, August 12th 1713, most humbly completed, the location of the salt springs, and strike of salt formations and rivulets in the same region of the Gommlo moorlands in the Wittenberg jurisdiction, according to the dowser from Freiberg, Johann Caspar Göbels, sketched onto this map, done in August, 1713."

The author of this statement and creator of the map was Johann Adam Schneider, a mine surveyor and second mine master (*Vice-Bergmeister*) at Schneeberg who had worked closely with a dowser from Freiberg. The record of their expedition—reports, expense accounts, letters—and two maps (one for the summer of 1713, one the summer of 1714) fill an official packet, and allows us to reconstruct much of these events. This becomes a short microhistory of prospecting for salt that introduces issues of central concern to all subsequent chapters. Mining officials like Schneider allowed that commoners who lived in the moor possessed important knowledge of the region, and they believed that dowsers could tap local knowledge and transmit it to them. Neither Schneider nor Nicolaus Voigtel, a second official who

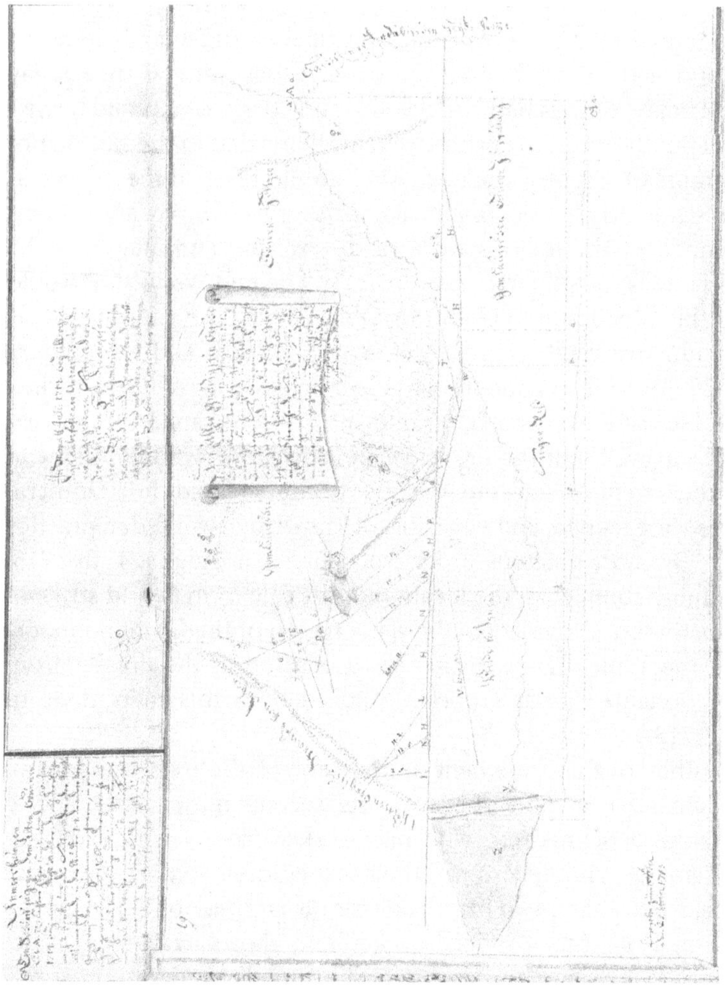

Fig. 2. Mine surveyor Schneider's first chart (1713) of Spitzberg Hill. A stream descends the hill and crosses the fields, dumping into a pond at left. Lines and coordinates mark the location of salt springs as proposed by the dowser.

oversaw these events, had a monopoly on geological knowledge here. The Freiberg Academy would advance geognosy, chemical mineralogy, and other earth sciences after 1765, and drilling technology would soon provide reliable information about salt formations. But drilling or boring was in its infancy in 1713, and even the Academy would neglect the subject of salt. Abraham Werner, the most famous geognost and mineralogist in Freiberg, said little about salt. While there was plenty written concerning salt and alchemy, only a few major texts treated the subject of salt production. Among the most important available in German to which Saxon officials could turn in 1713 were Georg Agricola's, *De re metallica*, Chapter Twelve (1556), Johann Thölde's (Basil Valentine), *Haligraphia* (1612), and Friedrich Hoffmann's *Beschreibung des Saltzwerks in Halle* (1708). The authors had much to say about the history of salt works, and the technologies of brine evaporation, but little on prospecting for mineral springs.[1]

While the officials quickly found themselves inundated with popular lore in Dübener Moor, much as their trial digs quickly filled with water, the men were never lost among unverified tales and claims, as if pre-modern mining operated in a scientific and technological vacuum. The officials had technique and knowledge, however premodern some of it may seem to us: they studied waters and sands, tasted residues, erected pipes and scaffolding, and trusted the dowsing rod. The hired dowsers would weave through the maze of tales, suggestive digs, and dead ends to provide Schneider with trustworthy information for his map. The dowsers enjoyed a high level of credibility in part because they had an association with Freiberg, to which mining officials increasingly turned for reliable geological knowledge. The Freiberg dowser was not simply a pre-modern contrast to the earth science specialist who brought the Academy international fame after 1765, but rather a true and immediate predecessor,

[1] Drilling for salt in Saxony began around 1714 and was instrumental in discovering rich springs after the 1720s under Johann Gottfried Borlach, though drilling was costly and dangerous, and both the drilling and geology of salt matured in the nineteenth century. Robert P. Multhauf, *Neptune's Gift: A Common History of Salt* (Baltimore: Johns Hopkins University Press, 1978), 178. On the science of salt and professionalization of salt production in the later eighteenth, and nineteenth centuries in Saxony and Austria, see Jakob Vogel, *Ein schillerndes Kristall: Eine Wissensgeschichte des Salzes zwischen Früher Neuzeit und Moderne* (Vienna: Böhlau Verlag, 2008), Chapter Two. See also Robert P. Multhauf, "Geology, Chemistry, and the Production of Common Salt," *Technology and Culture* 17:4 (Oct., 1976), 634–45.

as the dowser performed a similar function in mining as the scientist would.

Before introducing the cast of characters on this expedition to the Dübener Moor, we should recall the great importance of salt in the pre-modern world, and for Saxony in particular. Salt was a major sticking point between Saxony and Prussia in the decades leading up to the Seven Years War (1756–1763).

White Gold

There are good reasons why the word 'salary' derives from the Latin for 'salt' (*sal, salis*). Prior to mass production in the nineteenth-century, salt was one of the most important items on the planet. There were three major uses for this mineral: as a spice, as a preservative, and as medicine. And there were three major sources: the oceans, salt-rock veins, and natural springs or brine. Today, most of the world's salt comes from large salt-water operations and deep rock mining, and many natural springs that once fed the market flow untapped or serve spa resorts. This situation contrasts markedly with pre-modern continental Europe, when many territories like Saxony were land-locked, and deep mining in general was rare. Faced with the choice between costly long-distance trade and developing more local mineral springs, the territorial lords opted for the latter. Scores of towns and cities were founded as centers of salt making from the 12th century onward. The Celtic for 'place of salt' is 'Hall.' In Germany, Halle in present-day Saxony-Anhalt, Reichenhall in Bavaria, and Schwäbisch Hall in Württemberg, are among the many examples of towns that can trace their names back to the Celt. Other important late medieval German salt towns were Frankenhausen (today Bad Frankenhausen) in Thuringia, Kolberg in Pommern, and Lüneberg in Lower Saxony. The Tyrol too is rich in 'Halls' and salt mining history.[2]

Like mineral ore processing more broadly, salt mining was divided roughly into two stages: the digging and raising of the raw product (*Sole*=brine) in the field, and the more skilled work of extracting salt from the raw material in special pan houses. The various steps involved changed little from the late medieval period to the eighteenth century.

[2] Rudolf Palme, "Alpine Salt Mining in the Middle Ages," *Journal of European Economic History* 19 (1990), 117–136. Vogel, *Ein schillerndes Kristall.*

Having discovered a mineral spring either at the surface or some meters underground, workers had to isolate the mineral waters from mountain waters (*Wildwasser*='sweet' or 'wild water') with which it generally mixed. This required channeling the mountain waters away from the spring through wooden or iron pipes or conduits, and raising water in buckets or pumping machines. Once better isolated, the work of raising and channeling the brine itself could begin. The workers scooped the product out in buckets, filled barrels, and transported these to outdoor dugouts to dry in the sun, or to the salt works for boiling in pans [Fig. 3]. After the 1570s, one stop along the way might be the 'thorn house' (*Gradierwerk*), a large structure filled with straw, and

A—TRENCH. B—VAT INTO WHICH THE SALT WATER FLOWS. C—LADLE. D—SMALL BUCKET WITH POLE FASTENED INTO IT.

Fig. 3. An idealized image of salt workers collecting brine from a mineral spring for evaporation. From Georg Agricola's *De re metallica* (1556).

then blackthorn, designed to filter out more mountain water and increase the salt concentration of the brine. Then it was off to the pans as usual.

At the salt works or salines (*Siedehütte*), the waters were poured into large pans and placed over high-temperature fires. Once a saturation point was reached, the fires were lowered to allow a gradual crystallization to occur, and the formation of salt cakes. Workers raked off impurities as they surfaced, poured more salt water in, and produced ever-larger cakes. The product was shoveled into baskets, dried still further, and prepared for sale.[3] Only in the later eighteenth century, after coal-fueled fires, larger and stronger metal pans, and greater mechanization and bureaucratization, did salt processing, like the metals industry, undergo substantial changes.[4]

Like mineral earth, the springs underground were the property of the territorial lords, who loaned out the right to work them to private investors. These would have to organize should they protect their interests from both the lords and the workers they hired. The organization was called the panners collective (*Pfännerschaft*) after the German for 'pan' (*Pfanne*), and it was similar to the group of shareholders (*Gewerckschaft*) who financed hard mining. Unlike the shareholders, however, the panners collective was required to remain in the area of salt making. There were also no shares divided and sold (*Kuxe*) in salt mining, as required in hard mining. Finally, the princes of sixteenth-century Europe had already taken control over much of the hard mining, whereas in salt, the state took increasing control only in the eighteenth century. Salt extraction remained a more decentralized and low-cost enterprise until larger salt works and filtration houses became common.[5]

Focusing now on Prussia and Saxony, after 1680 the salt rich areas southwest of the Elbe all passed to Frederick William of Brandenburg-Prussia, as the panners collective began to yield before state power. His

---

[3] Hans-Heinz Emons and Hans-Henning Walter, *Alte Salinen in Mitteleuropa: zur Geschichte der Siedesalzerzeugung vom Mittelalter bis zur Gegenwart* (Leipzig: VEB Deutscher Verlag für Grundstoffindustrie, 1988), 16–26. Claus Priesner, *Das deutsche Salinenwesen im frühen 17. Jahrhundert* (Munich: R. Oldenbourg, 1980).

[4] Thomas Hellmuth and Ewald Hiebl, eds., *Kulturgeschichte des Salzes, 18. bis 20. Jahrhundert* (Vienna: Verlag für Geschichte und Politik, 2001). Multhauf, *Neptune's Gift*, Chapter Two.

[5] On salt and the state in Europe and abroad, see S.A.M. Adshead, *Salt and Civilization* (New York: St. Martin's Press, 1992).

successors developed Halle with new brine conduits, coal-fired furnaces, and large horse-powered pumps, and founded major new salines at Schönebeck. State interest in salt production in Saxony to the east lagged behind Brandenburg well into the eighteenth century. In fact, Saxony relied heavily on salt import from its salt-rich neighbor, providing wood for Halle's fires in return. This finally changed after the reign of Frederick August I ('the Strong'), Elector of Saxony after 1694 and King of Poland after 1697, whose mining experts did much to jumpstart the Saxon economy.[6]

One name worthy of note is Johann Christian Lehmann, who had defected from Prussia to work for Dresden. His text on boring for salt of 1714 was one of the earliest on the subject, and another work of 1721 argued forcibly that Saxony could compete with Prussian production. It had the following, rather straightforward title: *Saxony can develop usefully and productively all of the weak salt brine sources which it possesses in great numbers and can supply itself with salt within two or three years, no matter what others may say.* The Elector of Saxony gave Lehmann wide powers in 1717 to explore and open salines anywhere in the realm. A second important name in the salt history of Saxony is Lehmann's former assistant, Johann Gottfried Borlach (1687–1768), who worked for the Elector both in Saxony and Poland. Borlach had already discovered mineral springs of high salt content, and had been searching for some twenty years at Dürrenberg, when he broke into a rich spring there in 1763. At a depth of 223 meters, Borlach's diggers tapped mineral waters containing some 10% salt. Construction of salt works began quickly, and soon Saxony was cutting back on its Halle import considerably. By the end of the century, there were 11 salt houses and 30 large metal pans in operation at Dürrenberg. Two other major salt towns to emerge under Borlach's direction were Artern (~1725) and Kösen (~1730).[7]

Most mineral springs did not produce the quantity and quality of Dürrenberg in Saxony or Halle in Brandenburg. A more typical example might be Auleben near the Harz Mountains (Thuringia today).

---

[6] Paul Reinhard Beierlein, *Neues über die ältere Geschichte der Salzquellen zu Altensalz* (Berlin: Akademie Verlag, 1962). Manfred Straube, "Soziale Lebensbedingungen und soziale Sicherheiten von Arbeitern in der pfännerschaftlichen Saline Halle a.d. Saale in der zweiten Hälfte des 18. Jahrhunderts," in Werner Freitag and Heiner Lück, eds., *Halle und das Salz: Eine Salzstadt in Mittelalter und Früher Neuzeit* (Halle: Mitteldeutscher Verlag, 2002). Multhauf, *Neptune's Gift*, 83–88.

[7] Emons and Walter, 56. Multhauf, *Neptune's Gift*, 88–92.

The locals had known since ancient times that a spring lay near the village. The first organized effort to tap the site and construct salines dated to the sixteenth century, especially during Elector August I's reign, when Dresden ruled lands this far west. Undeterred by claims that the waters were weak in salt content (less than 1%), and that the spring was unstable, changing position, the Elector sent some 100 workers to Auleben to realize his plans. By 1568, he had two houses in operation, and the dig reached some 36 meters. But the product never offset the cost of labor and production. The foreman eventually fled, fearing the wrath of his powerful patron, and in 1572, the Elector finally released the workers and closed down the works. Others would attempt to revive the Auleben site over the decades, but with little success. A commission in 1675 found only abandoned works, and today there is little to remind the passerby of any formal interest.[8] Much the same story could be told of Altensalz in the Vogtland, and any number of smaller springs strewn across Saxony.[9]

The expedition for salt in Dübener Moor between 1713 and 1714 dated to the reign of August the Strong, just prior to Borlach and the first application of drilling technology, and it was no more successful than Auleben. It was one of many ventures that Dresden authorized throughout the century that produced no lasting results. There was nothing extraordinary about it: no unusual discovery, no implementation of new technology, no famous scholar or bureaucrat involved like Lehmann or Borlach; although Nicolas Voigtel, overseer of mines (*Bergvogt*) and mine surveyor to Thuringia (part of Saxony), had achieved some fame for his surveying manual. The expedition Voigtel headed was not an outright failure, but an example of the more everyday successes and failures that characterized the majority of expeditions and exploratory digs throughout the century. Ordinariness in this case is no disadvantage for us, since we might assume that the knowledge and skill it featured was representative for the period, or at least more representative than Borlach's famous expeditions with drill bits. Then again, microhistories often demonstrate how extraordinary the ordinary truly is: we will witness collapsing digs, a worker who almost cut off his foot, a peasant's strange tale of springs, and dowsers proposing numerous digs.

---

[8] Emons and Walter, *Alte Salinen*, 89–90.
[9] Beierlein, *Neues über die ältere Geschichte der Salzquellen*.

A Tale of Salt

The expedition would take place near the town of Gommlo along the northeastern border of the Dübener Moor in present-day Saxony-Anhalt. The place lay, according to one of our actors, "on the road between Düben and Wittenberg."[10] In the summer of 1713, a small party set out for the 122-meter high Spitzberg Hill, rumored to contain rich untapped brine. The Dübener Moor, which lies between the Elbe and Mulda rivers east to west, and between Leipzig and Wittenberg north to south, is one of the largest forested regions in Germany. The physical geography of the moor consists of sand crisscrossed with countless swamps, springs, and creeks, rolling hills and occasional small mountains like Spitzberg, all originally plotted by receding glaciers of the Pleistocene periods. It is home to any number of spa towns today, including Bad Schmiedeberg and Bad Düben. The area near Gommlo lay in the Wittenberg jurisdiction (Amt Wittenberg) in 1713, an administrative unit of Saxony. Had the expedition succeeded in locating new sources, the Elector would have founded salt works to help compete with Halle and other Prussian centers. This was not the first time that an expedition from Dresden came to Gommlo with that hope, as we shall see. But Saxony would not become competitive before Borlach, and there would never be a 'Bad Spitzberg.'

A few important features about prospecting for brine will emerge in this expedition: most notably, the dowsing rod was the chief prospecting tool, as dowsing and mine surveying appear as complementary practices. Two dowsers from Freiberg largely determined the course of the expedition, as they paced the field before officials and diggers eager to hear of promising spots. There was no mention of the derogatory 'wishing' or 'divining' rod (*Wünschelrute*), but always the more commonplace 'rod' (*Rute*), which suggests a high level of familiarity. Its legitimacy never came into question, although the claims of one practitioner, a peasant of uncertain origins, were treated with suspicion, suggesting that credibility rested on experience in mining and association with Freiberg. Second, we see how dependent the search for salt was on the memories and experiences of local inhabitants in the marshlands. The officials welcomed tales of salt that surfaced from the commoners, even as the men maintained strong top-down control and

[10] Sächisches Bergarchiv Freiberg [BAF], Loc. 40013, Nr. 174, 21b.

directed all labor. Their trusted dowsers would make sense of it all. Prospecting practice involved a synthesis of high and low cultures and traditions. Finally, and more fundamentally, this was a messy business. Searching for mineral waters in pre-industrial Europe was an ongoing battle against wet sands, mountain waters, and swamps. A dig continually threatened to collapse, adequate equipment and supplies were not always forthcoming, and the wellbeing of the workers and dowsers was always in jeopardy. The officials were flexible and responded to changing circumstances, but they were also rational accountants directly responsible to Dresden. Decisions had to be made and expenses tallied: our men considered all available information, from the quality of the earths and waters discovered, to the statements of dowsers and local inhabitants familiar with the terrain. Drawing no modern distinction between rational and irrational methods, the mining officials were perfectly capable of imagining mechanical solutions to problems—a water wheel for drainage—as entertaining the quasi-magical solution of dowsing.

Beginning now with the central tale that motivated the expedition, we see that stories of hidden springs and streams in the Dübener Moor had been told for centuries. According to a city council representative (*Amtsverwalter*) at Wittenberg, Samuel Francke, a particular area near the town of Gommlo was mined for salt in 1575, and again in 1686, but has been dormant ever since. As early as 1709, Francke had been writing Dresden with a view to generating formal interest. In a letter of 1711, we learn that the head forester (*Oberförster*) to Pratau, just south of Wittenberg, had recently come to Francke with a 28-year old farmer named Hans Fischer. This man, who would join the expedition, told a tale that later folklorists like Herder and the Grimm Brothers would have found most endearing:

Fischer's father-in-law knew a shepherd, or so it went, who, some 50 years prior, had discovered a curious stream not far from where his livestock roamed. The shepherd bottled up a sample and returned home. He boiled it in his kettle to study the residue, and he marveled over how rich the salt content was. Salt workers from Halle (*Halloren*) had heard of this promising discovery, and so they came looking for the shepherd, who promptly led them to the spot. The men, anxious to keep the find a secret before returning from Halle with more supplies, plugged up the spring with stones, bits of clothing (*Sammetärmel*), and pig fat (*Schweinschmer*). They then left for home, bringing the poor shepherd along. "He was never seen again," said Fischer's father-in-law to Fischer. The old man kept the tale secret, until, in his final years (apparently almost 100), he unburdened himself to his son-in-law, and wished to show him the spot. But Fischer did not indulge his aging father-in-law before he passed away. Now the good peasant, perhaps a little remorseful, claimed to know the spot, and had already shown it to the head forester.[11] It lay on Spitzberg Hill.

That was 1711. That men from Halle would be interested in a potentially competitive spring in Saxony is certainly probable, even if we smirk at the details of Fischer's account. But tales alone would not win support from Dresden. Francke's letters did not spark any official interest before the summer of 1713, after he had invited a mine master to inspect the area. Johann Meinel from Glashütte in the Erzgebirge came to the Dübener Moor in June to confirm Francke's suspicions and Fischer's story. Francke described Meinel as a man of great reputation for his "miner wisdom" (*Bergmännische Wissenschaft*).[12] As in the mining books we review in Chapter Three, '*Wissenschaft*' referred here to great experience or wisdom, rather than theoretical knowledge or university training. According to Francke's report to Dresden of his encounter with Meinel, he had studied the area that Fischer had specified, and agreed that untapped salt springs lay somewhere below.

Possessing miner wisdom in 1713 also meant understanding the secrets of dowsing, since the strong response of Meinel's rod provided additional evidence for his claim. It dipped above Fischer's spot, "like nowhere else in this entire region."[13] In fact, Meinel prepared his own report for Dresden, in which he substantiated Francke's claim and commented on Meinel's own dowsing. His remark indicates that, however contemporaries explained the functioning of the rod in mining, those explanations generally concerned mineral ores, not springs. Meinel wished to extend dowsing theory to include brine. He explained that salt springs had a particular direction and angle (*Streich*), just like mineral veins and rocks did, to which the rod dipped. Like mineral earth, salt occurred in distinct enough formations of brine or rock for the dowsing rod to detect. Meinel did not explain the actual mechanism at work—he was assuming some sort of affinity or

[11] BAF, Loc. 40013, Nr. 174, 10–11.

[12] "...wegen seiner Bergmänischen Wissenschafften in sonderl. guten Ruff stehet." BAF, loc. 40013, Nr. 174, 7.

[13] BAF, Loc. 40013, Nr. 174, 7b.

emanation from the formations—but simply referred to the strike of the salt, as if to remind the reader that it was the strike of mineral ores that the dowsers followed above ground, and so it stood to reason that a skilled dowser could follow salt sources as well. Meinel need not have worried. Indeed, the language these men used to refer to salt springs—'vein' 'strike'—was drawn directly from ore mining, and the applicability of dowsing in salt prospecting was never questioned in this case. Meinel recommended reopening and deepening the abandoned salt mines of Dübener Moor.[14]

Francke and Fischer's hopes were finally realized in early August of 1713. A formal expedition was planned for late August with mining specialists and a hired Freiberg dowser, who, it was expected, would check Meinel's pronouncements. Freiberg emerges as the site for the most credible and best-known dowsers. A privy counselor in Dresden, Baron Johann Egidius von Alemann, wrote to Francke expressing the Elector's approval of the expedition. Both Alemann and Johann Gottlieb Voigt of the Saxon Mining Office (*Oberbergamt*), Freiberg, wrote to Schneider, mine master and surveyor at Schneeberg, calling him to the Dübener Moor to lead the expedition. He was to, "see the council representative [Francke], inspect the claim, and also have someone examine with the rod."[15] A Dresden bureaucrat and nobleman could speak casually about the 'rod' (*Ruthe*). Schneider should then prepare a map (*Riss*) of the dowser's pronouncements. The practitioner in question was Johann Casper Göbels "from Freiberg," about whom we learn little beyond his hometown, Erbisdorf, just south of Freiberg.[16]

The results of Göbels's dowsing with Francke, Meinel, and Schneider are given in Schneider's map [Fig. 2]. Turning to this drawing, we see a pond at the far left, across which the road from Bad Düben to Wittenberg ran. We see numerous crisscrossing lines, along which Schreiber has indicated the strike in degrees (according to the compass). These are Göbels's many claims, as according to the dowsing rod—far more than discussed in the reports. At center there is the scroll with title, and not far from the lower left-hand corner of this scroll we can find the letter 'A' on the map, just beside a circular,

---

[14] BAF, Loc. 40013, Nr. 174, 11b–12b.

[15] "...in Augenschein nehmen, auch mit der Ruthe der Gegend einen Versuch thun laßen." BAF, Loc. 40013, Nr. 174, 1.

[16] BAF, Loc. 40013, Nr. 174, 14.

darkened formation. According to the notes (*Anmerckungen*) that surround the drawing, this was the salt spring worked in 1575 under August I, abandoned, reopened in 1686 under Johann Georg III, and abandoned again. The dowsers pronounced that two "seams or rivulets" (*Klüffte oder Ader*), labeled 'B' and 'C' on the map, intersected by the old spring. The two oblong formations to the left of the spring were swampy grounds. If we follow the spring's flow downhill (also darkened), we hit a narrower creek at 'W.' This was the site of abandoned piping, once used to separate mountain waters from the brine. The long, narrow creek traverses the entire map and runs into the pond at left. If we follow this creek to the extreme right, we hit another stream uphill, and at that intersection Schreiber has placed 'A a.' This would be the site of piping works constructed by the present party the following summer.

But let us follow the events as they unfolded, beginning with two letters that commented on this preliminary expedition with Göbels. The first is a letter from Francke to the Freiberg Mining Office, dated October 19, 1713. The city council representative reported that, while Göbels did indicate rich salt springs, he did not confirm Meinel's pronouncements exactly. Göbels repeatedly passed over the spot on which Meinel had experienced a very strong dipping motion. Some ten 'steps' (*Schritte*) away, however, Göbels's rod dipped with the same force as Meinel's had. Assuming an adult step to cover a little more than one meter, ten steps may have amounted to ten or eleven meters.[17] The second letter is a more comprehensive summation by Schneider to Dresden dated October 26. He explained how, in the company of Francke, he examined the entire region in question with the "experienced" dowser from Freiberg.[18] Like Francke, Schneider noted that the men had not discovered a new salt spring, but he stressed that Göbels's and Meinel's pronouncements otherwise "agreed," and that the underground rivulets they detected were "especially strong." The author recommended that digging continue in search of the source of these rivulets.[19]

These letters and reports from the summer and fall of 1713 end abruptly at this point. Aside from one letter dated December 1713

[17] BAF, Loc. 40013, Nr. 174, 21.

[18] "Die gantze Gegend… mit der Ruthen durch einen geübter Ruthen Ganger alles Fleißes untersuchet." BAF, Loc. 40013, Nr. 174, 21b.

[19] BAF, Loc. 40013, Nr. 174, 22–23.

authorizing a second expedition, the material that follows concerns the summer of 1714. It is likely that expeditions such as these halted during the winter months, when the hard ground and snows made it difficult for miners and prospectors. The search now featured an even higher-status official, the head of mining (*Bergvogt*) for Thuringia, and mine surveyor, Nicolaus Voigtel. Dresden called for "one or two good dowsers, as well as skilled workers" to work with Voigtel and other officials.[20] The dowsers, we see, were distinguished from the diggers. The men of pick and axe (*Hauer*), while they enjoyed special privileges, were laborers, not unlike the woodcutters we will also meet; the dowsers worked more directly with mine surveyors, mine managers, and other higher-status officials from mining centers like Schneeberg and Freiberg.

The higher status of the dowsers from Freiberg vis-à-vis the laborers is evident in the expense accounts that now appear, as the expedition became more formal and expensive. One representative accounting was drawn on July 14, 1714. The peasant, Hans Fischer, was now given 10 groschen to show the party his supposed salt spring, supposedly hidden by Halle salt workers; a digger had a weekly salary of 1 taler and 9 groschen; a foreman earned 2 taler per week; whereas the dowser, Göbels, earned 7 taler and 19 groschen for travel expenses and 20 days of work, which was slightly higher than the foreman's pay. The tally specifies Göbels's work: "to dowse with the rod" (*mit der Ruthen zu gehen*), and there is no indication that he did any manual labor or other work. A second dowser from Freiberg, one Christian Eickern, earned 6 taler and 4 groschen for travel expenses and 13 days of work. Of course, the dowsers earned far less than Schneider and Voigtel did: 38 taler and 14 groschen for 28 days, and 38 taler for 28 days, respectively. The hired dowser from Freiberg was no overseer, mine master, or surveyor, but he was no peasant or digger either, men who labored with their hands.

Otherwise, the list includes a host of tools and other supplies involved in prospecting for salt springs on a small scale, providing us with good information on what this work entailed. The total cost for supplies for the two-week venture came to 39 taler, 10 groschen, and

---

[20] BAF, Loc. 40013, Nr. 174 [no page number]. The transition from 1713 to 1714 also marks the end of page numbering (at 23), just prior to this letter. In what follows, I provide dates and signatures as best I can, and impose a numbering system for one longer report titled, 'Actum, Göbels Haus.'

10 pfennig. Some of the tools were involved in basic digging, such as an iron shovel, hammers, chisels, a scraper, and axe. A certain "large hammer" alone cost over 1 taler. Other tools would have been used to shore up sand and waters inside a dig, such as saws and nails for wooden boarding. The wheelbarrow and bucket were obviously intended to move items around. It is interesting that the cost of transporting and employing the dowsers appears at the bottom of this list of supplies, rather than elsewhere. The dowsing rod and its bearer was essentially another tool, although clearly separated at bottom as a major expense. Voigtel and Schneider's salaries appear in these accountings further removed from the list of tools.[21]

'Actum, Göbels Haus'

More exact particulars concerning this expedition appear in a twenty-page report by Schneider and Voigtel. Born in Freiberg, Nicolaus Voigtel (1658–1714) moved with his father, a miner, to Mansfeld in Thuringia. The son rose to become head of mining in Thuringia and tithe-collector (*Bergzehender*) to Mansfeld.[22] Among other major projects, he once sought to expand the salt production of the land. He proposed a 50-meter long combined graduation house and saline that would cut down on wood usage by using the same fuel source for both processes.[23] Voigtel's *Subterranean Geometry, or Mine Surveying* (*Geometria Subterranea, oder Marckscheide-Kunst*) went through numerous editions after 1686, including a 1713 publication, not long before the present expedition. The text includes one of the earliest explanations of trigonometry for miners, and books on mining and metallurgy after Voigtel frequently referenced his work.

While Voigtel's manual provided no analysis of dowsing, he mentioned the practice in the same matter-of-fact way he would deal with dowsers in the field. *Geometria Subterranea* focuses on the mathematics and instruments of surveying: lessons on arithmetic, geometry, and trigonometry; how to use an eyepiece (*Vizier*); how to use the compass to determine the strike of veins; and how to prepare a map (*Riss*) of

[21] BAF, Loc. 40013, Nr. 174, receipt for August 13, 1714.

[22] Allegemeine Deutsche Biographie, s.v. 'Voigtel, Nikolaus.'

[23] Uwe Meißner, "Die Gründung und der Ausbau der königlich-preußischen Salinen Schönbeck und Halle am Anfang des 18. Jahrhunderts," in Freitag and Lück, 86

veins. There is no dowsing rod among the many instruments and tools that Voigtel depicted in the title piece—measuring staffs, rope, and protractors. But when Voigtel turned to the practice of rendering mineral veins on a mining chart, and actually determining where to dig, we discover how closely the mine surveyor worked with dowsers. Surveying could be highly abstract, a matter of trigonometry and exact measurements, but the business of mining demanded more concrete information too. "It is often the case," explained Voigtel, "that before the miners can dig, or tunneling started, all the veins must be reckoned by the mine surveyor and dowser."[24] The miners needed to know if the veins would deepen, cross, or otherwise change course. Voigtel directed the reader's attention to an illustration of mineral veins, in which a dotted line marked the trajectory of the dowser's pace [Fig. 4]. "I let the dowser walk ahead," Voigtel said, and wherever the rod dipped, he struck a post into the earth, attached a cord that hung from a prior post, hung the mining compass, and used measuring instruments and trigonometry to determine angles and distances. The dowser then continued ahead, and the process began anew, as the dowser and surveyor worked in harmony.[25]

Other practical uses Voigtel found for dowsers included double-checking the accuracy of posts already positioned by the mine surveyor, and predicting where to dig a shaft that would meet an underground tunnel. If the determinations of the dowser and surveyor mostly agreed, "one is all the more certain."[26] If the men did not agree, then Voigtel recommended making a few test digs. A high valuation of empirical evidence did not disqualify the author's commitment to dowsing, as we might suppose today, assuming a fundamental incompatibility between knowledge claims derived from dowsing and those based on applied mathematics.

That Voigtel was an attractive choice among Dresden bureaucrats to lead the salt expedition of 1714 is clearest in chapters devoted to the subject of mineral springs. Voigtel was an expert on water management, piping, and digging in the most difficult of working conditions. Pipes of iron and lead were preferable, he argued, but also costly, so

[24] Nicolaus Voigtel, *Geometria Subterranea, oder Marckscheide-Kunst, darinnen gelehret wird wie auff begwercken alle Klüffte und Gänge in Grund und am Tag gebracht, auch solche von einander unterschieden werden sollen* (Eisleben, 1686), 81.

[25] Voigtel, *Geometria Subterranea*, 83.

[26] Voigtel, *Geometria Subterranea* (Eisleben, 1713), 140.

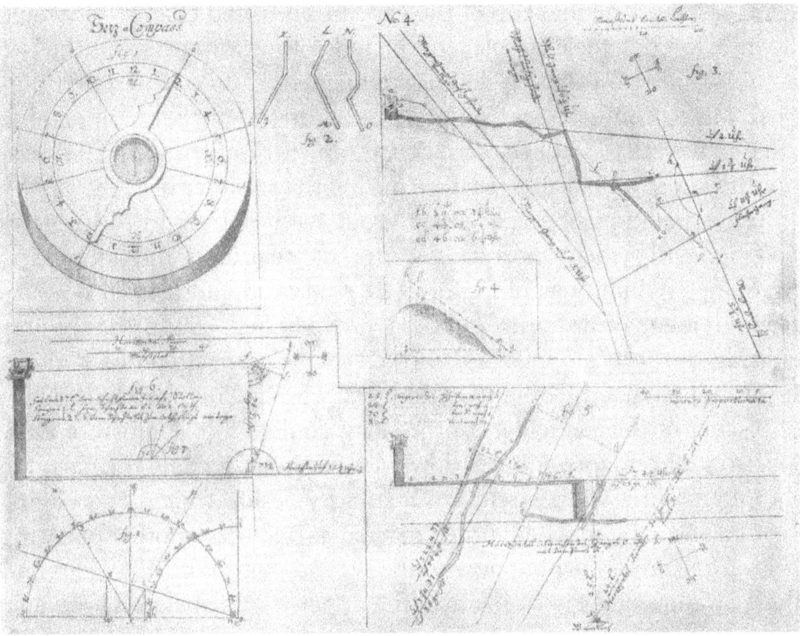

Fig. 4. Diagrams from Nicolaus Voigtel's mine surveying manual, *Geometria subterranea* (1686). The mining compass is shown at the upper left. A dotted curved line and proposed claims 'h' 'i' and 'o' at the upper right show the dowser's work.

Voigtel discussed the best sort of wood and how to maximize the force of gravity in moving mountain waters and brine.

Reading the title of Voigtel's report, 'Actum, Göbels Haus,' we see immediately how central the dowser was to mine surveying, just as the author had described in his *Geometria Subterranea*.[27] Not only would Göbels's practice steer the decision-making of the officials in charge, but also the party may have actually stayed with Göbels. The men gathered on June 28, 1714. The party consisted of eight men: Voigtel and Schneider, the two officials in charge; the two dowsers from Freiberg, Göbels and Eickern; a foreman named Johann Georg Böhm, and a digger, Johann Christoph Richter, both men from Marienberg; the head forester to Pratau; and lastly, the peasant and amateur dowser,

[27] 'Actum Göbels Haus.' This Actum, or report begins about halfway into packet Nr. 174, just after the second large (folded) map, or *Riss*. Pagination (1–10) is now internal to this Actum.

Hans Fischer. The makeup of the expedition would change—it would exclude Fischer, for example, and include other workers—but the two authors and two Freiberg dowsers remained its core. The stated purpose was to continue searching on and around Spitzberg Hill, and to attempt to verify Fischer's claim concerning the hidden spring. His was not the last story of salt to surface during the summer of 1714.

The expedition began by setting out with Fischer to the edge of Spitzberg Hill, where we quickly learn that the peasant did not possess the same credibility as more formally employed and trusted dowsers from Freiberg. Near the area under which Fischer's salt spring supposedly lay, the party stopped and allowed the "peasant" to proceed alone. It was hoped that the other two dowsers, who had no prior knowledge of Fischer or the particulars of the story, could either confirm or deny his pronouncements. Fischer dowsed a full hour, leading the party up the mountain a short stretch to a stream, beside which lay swampy ground and a dig that he claimed to have started with another party the prior Christmas. Turning now to the second map [Fig. 5], we can find these swamps and Fischer's dig at the upper-right, labeled 'A.' Adverse weather and temperature had forced him to abandon the site. It was the place where Fischer said men from Halle once plugged or redirected the waters with large boulders and other materials.

According to the authors, there indeed lay a number of large, rounded stones by Fischer's dig,[28] but there was no reason to believe that they had been purposely placed, and there was no cloth or fatty substance in the ground. There was also no indication that the waters here were salty. But Fischer was adamant, and when put on the defensive, he appealed to the "common talk" of local salt workers as well as to the sharp dip of his rod.[29] He wished to substantiate his claim by determining the strike of the salt waters below. But Voigtel and Schneider were losing faith in Fischer. His dowsing was unsound, or "had already been proven false" (when the Freiberg dowsers declared the waters 'sweet'), and Fischer "did not agree with our dowsers in the slightest." The decision was made to have Göbels and Eickern dowse the area more thoroughly, since only they would, "get to the bottom of the matter."[30]

---

[28] Wacke=large rounded boulder. See Jacob Grimm and Wilhelm Grimm, *Deutsches Wörterbuch* (Leipzig, 1854–1960), s.v., 'Wacke.'

[29] BAF, Loc. 40013, Nr. 174, Actum, 2.

[30] "Weiln mann vorher ihme albereit in seinen Ruthen gehen gantz falsch befunden, und auch hernach fast in geringster nicht mit unsern Ruthen Gänger über ein kam…

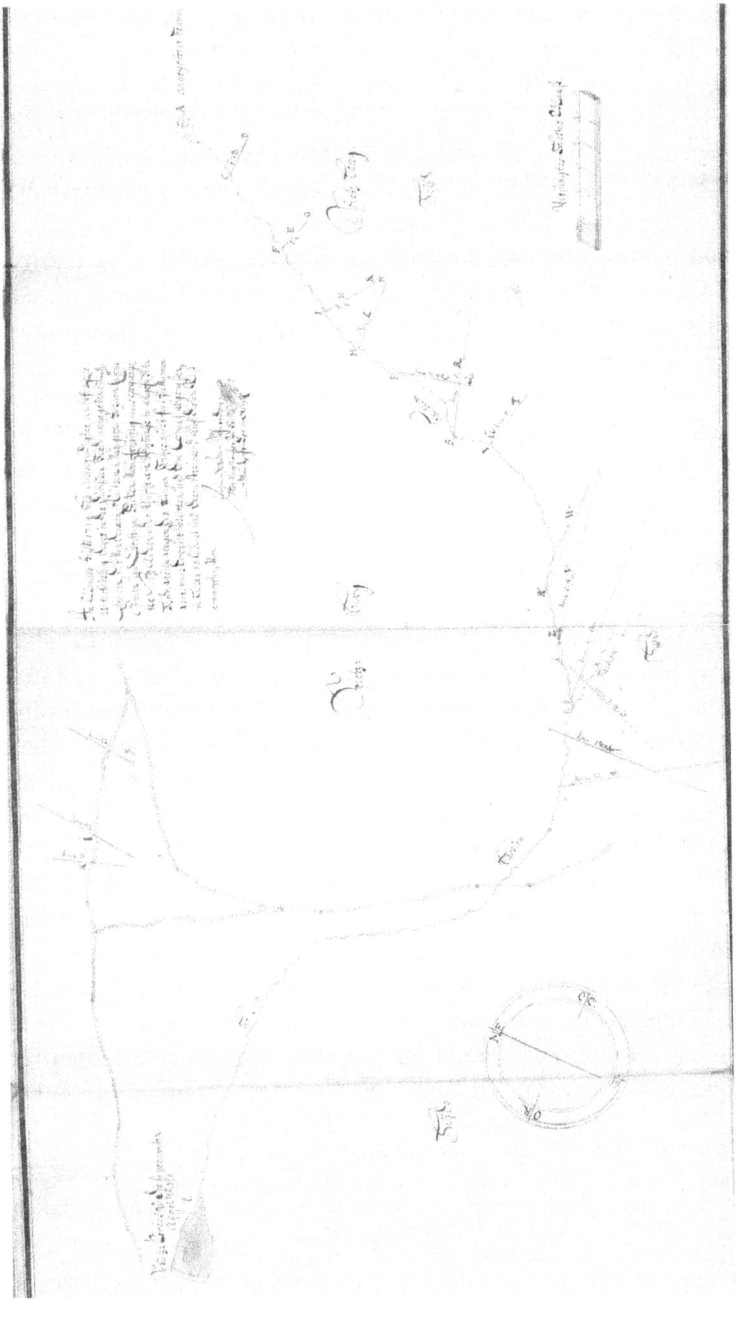

Fig. 5. Mine surveyor Schneider's second chart (1714) of Spitzberg Hill. A stream runs from high in the hills and dumps into the pond at far left, and lines and coordinates mark the location of salt springs as proposed by the dowser.

It is striking how little credibility the party gave to Fischer and his story after finding his dig. The men from Freiberg were "our" dowsers, whereas Fischer was merely a "peasant," as they repeatedly referred to him. We might expect that the officials began to lose faith in dowsing after this confrontation with Fischer, but nothing would be further from the truth: Voigtel and Schneider became ever more committed to their own trusted and handpicked practitioners. The rod itself was not on trial here, but rather Fischer's practice in particular; and the peasant with bold claims was not the mining officials' model of a trusted dowser. Long-standing reputation and association with mining, and a touch of humility made the legitimate dowser, as we shall see more clearly in Chapter Four.

The remainder of the report centered on Göbels and Eickern's pronouncements. First, Eickern set out with his rod to double-check the accuracy of Göbels's claims at the old spring, as recorded in the first chart of 1713. In marked contrast to the prior evaluation of Fischer, here, Eickern's work "agreed with Göbels's claims in everything, except that the men differed with respect to the quality of the waters."[31] In other words, Eickern's rod dipped at most of the same spots as Göbels's had, even though the men had different interpretations as to what exactly lay below. The implication was that such crosschecking mutually certified both dowsers, notwithstanding differences on more subtle determinations. They mutually discredited Fischer again when they checked his claim that strongly flowing waters lay just above the pond. Both of "our" dowsers, wrote the authors, found the waters to be rather weak.[32] The dowsers continued to work together, complementing each other's practice, as they studied the rivulets that traversed the old spring and swamps nearby. Since, "according to the rod," the waters were salt-rich, the party decided to undertake its first dig, labeled 'A' [Fig. 5]. On June 30, 1714, two diggers battled heavy, wet sands for 1.5 Lachter (3 meters) before it was concluded that without more diggers and adequate equipment—wood for support, pales, shovels—the digging would have to cease. "The sand just would not hold."[33]

so hat mann darauf wenig bauen können, sondern diese ganze Gegend durch unsern Ruthen Gänger umb genauer hinter die Sache zu kommen, aus refiren zu lassen beschlossen." BAF, Loc. 40013, Nr. 174, Actum, 2–2b.

[31] BAF, Loc. 40013, Nr. 174, Actum, 2b.

[32] "Uber den Teich gab der Bauer ingl. eine mächtige Ader an, unsere beyden Ruthen Gänger aber fanden solche sehr schwach." BAF, Loc. 40013, Nr. 174, Actum, 3.

[33] BAF, Loc. 40013, Nr. 174, Actum, 4.

The dowsers were asked to continue pacing the moorland, but with one caveat: noting the cost of the expedition, the authors requested that the dowsers "consider all water sources, vitriolic and aluminic, as well as salt and wild [sweet]." In other words, Voigtel and Schreiber wished to pursue *any* lead, and they did not wish that their dowsers searched only for pure salt (NaCl) to the exclusion of other salts and minerals. But the dowsers insisted that their rods were dipping to purer salt, not vitriolic or aluminic water, even though a vitriolic taste had indeed been determined.[34] Were the officials suggesting that the dowsing rod obeyed the desires and intentions of its bearers? Were the dowsers maintaining that the connection between subterranean formations and dowsing rods was more physical or mechanical than mental? We cannot say, but the officials will repeat this request that the dowsers focus their minds in a particular way. Voigtel and Schreiber, now conducting as broad a search as possible, took a sample from the old spring, "as deep as possible," or as untainted by mountain waters as possible, and boiled it to determine any salt content. The residue showed no salt. It appears from later entries again that Göbels's house had doubled as a makeshift laboratory as well as residence, since the party was using his kettle to boil and test samples they took from the mountain.

On June 29th, Göbels and Eickern had also dowsed the area on the Spitzberg by Fischer's dig, following the stream downhill. Schreiber used the compass to take the direction and angle (strike) of the dowsers' findings, but apparently the bush was so thick, or growth so twisted, that stretching the cord (*Schnur*) to take exact measurements and angles was a problem, and the surveyor was forced to reckon with steps. This is a useful reminder of the difficulties of using geometry and trigonometry in the field, and helps explain why a trusty dowser worked alongside a surveyor. Schreiber may have turned to his dowsers when he could not locate a distant point through the trees, lay his cord taut through the bush, or otherwise rely on his instruments in harsh conditions. Some Lachter downstream, the dowsers detected a "rather noble salt vein" that crossed the stream (labeled 'B' with ends 'C' and 'D' on Fig. 5). Further salt formations seemed to cross the stream as the men walked downhill, until they arrived at 'Z.' This was a swampy area

[34] "...sie bleiben aber, ob gleich was vitriolisch biß weile zu spühren, dabey, das ihre Ruthen auff ergiebige Saltz Adern zeigeten." BAF, Loc. 40013, Nr. 174, Actum, 4.

that was decidedly reddish, and in which the stones carried a reddish sedimentation or deposit (*Sinter*). The men, suspecting either vitriol or iron to be the cause of the reddish appearance, wished to construct piping at this site to channel off the mountain waters and isolate any brine.[35] This would require additional equipment—shovels, pales, pipes—and it seemed that an order from Wittenberg had been delayed.

It was now July 1, just three days into the expedition. The men dispatched a man to Wittenberg to pick up any equipment that was complete, but learned that only some "old" iron pipes and suction pumps (*Pumpen*) were ready. The authors may have meant that the equipment was rusted and weak, since they later placed an order in another town for material that was, "stronger than the Wittenberg."[36] They also mentioned that they did not need the first order "yet," which would imply that they wished to establish a trial dig some Lachter deep before employing expensive equipment. Meanwhile, Göbels had become ill and remained at home. The circumstances behind this illness were not explained, though he did perform again, so his condition must not have been too grave. The dowsing continued unabated, first by both dowsers in the immediate vicinity of Göbels's place, and then by Eickern alone near the old spring (shown on Fig. 2). The men detected no salt near the residence, but Eickern did have a powerful (*heftig*) reaction in a wooded region some Lachter downhill from the spring (at the crescent moon shape, just below the scroll). The waters at the dig beside the spring, where the "sands would not hold" on June 30, may have continued downhill and fed a larger underground stream or spring at the present spot. This was the explanation provided for Eickern's strong reaction here. On July 3 a dig was ordered.

The results were highly suggestive: at around one Lachter (2 meters) deep, the diggers uncovered "cloudy" water that did not clear after sitting undisturbed (indicating a highly soluble substance, or salt). The party suspected a form of salt, but after boiling a sample, this proved untrue. The digging then continued another few meters, until the workers came across a weak spring that flowed a few meters before disappearing into mountain waters. But now the work had to be discontinued again, and rudimentary wooden supports abandoned, due

[35] Röschen= to channel or trench. BAF, Loc. 40013, Nr. 174, Actum, 3b.
[36] BAF, Loc. 40013, Nr. 174, Actum, 5b.

to the unrelenting mountain waters and lack of men and equipment. The foreman from Marienberg, Böhm, was sent to Schmiedeberg (today, Bad Schmiedeberg) to question its craftsmen about equipment, but he learned that they had nothing prepared. Böhm placed a short order for the following day. Digging here would resume a few days later.

Back on Spitzberg, where the party had discovered the reddish swamps, the two diggers were channeling off mountain waters, and some supplies finally arrived on July 5 from the Wittenberg order. Progress was slow, because the diggers had to chop and gather wood to construct supports. More samples were taken, still without salt, although a second rumor of salt did surface. The diggers, doubling as woodcutters, had to deal with the Klaftermeister to Selchau (possibly Sachau, just east of Schmiedeberg), whose job entailed apportioning bundles of wood,[37] and who it seems had his own experience searching for brine on Spitzberg. He claimed to have personally discovered waters that, after boiling, had left a grayish salt behind. Encouraged by this new information, Voigtel and Schneider personally set out to Schmiedeberg that evening to pick up the latest order and place another. None of the equipment was ready, they learned, so the men decided to spend the night, in the expectation that the work would be completed the following day.

On July 6, the officials sent one digger to Selchau to speak with the Klaftermeister again and encourage him to return to the Spitzberg to identify the spot for the party. He would have to decline this request, he said, because his superior, the Wildmeister,[38] would not grant the absence. But the Klaftermeister insisted that his salt source existed, and that a grayish substance appeared in his kettle after boiling. The powder had a salty taste, though he could not swear that it was salt. Meanwhile, the completed order from the Schmiedeberg artisans arrived that evening, and Voigtel and Schneider could plan to return to Göbels's place the following day.

Early on the 7th, the men arrived at Göbels's to find that the miners had boiled further samples from the Spitzberg dig, and that the brownish residue had a distinctively salty taste. The officials immediately sent for more samples, and again, the same brownish, salty residue appeared.

---

[37] See Grimm and Grimm, *Deutsches Wörterbuch*, s.v. 'Klafter.'

[38] An overseer of hunting, and increasingly woods as well. See Grimm and Grimm, *Deutsches Wörterbuch*, s.v., 'Wildmeister.'

Perhaps the Klaftermeister meant this very swamp, they reasoned, and ordered that channeling continue. Voigtel and Schneider, obviously quite experienced with the properties and behavior of mineral waters, noticed that the salty residue did not truly dissolve into the waters from which it came. The waters were brownish, but did not fully absorb the salt. The men concluded that the substance could have been chafed off from mineral rock through the action of flowing waters (*Auslaugung*, lixiviation), rather than be part of a brine solution.[39]

The next encouraging news came on July 9 when the diggers, now using more adequate tools and wooden conduits, noticed chunks of mineral rock (*Geschiebe*) in the sandy waters along the channel. This would support the notion that the salty waters just discovered were the product of lixiviation, as the fragments suggested a nearby hard vein. Such alluvial fragments were well known to experienced diggers: they studied the shape and size of each rock, since it was believed that these qualities helped determine the relative distance from the mineral veins from which they originated. The smoother and smaller the surface, the further the stone had traveled from its place of origin, and the longer the waters had weathered it down. The digger might even claim to know the direction to the mineral vein as based on the position of the fragments in the earth. According to the report, the party found two sorts of fragments that day on the Spitzberg: sulfuric and aluminic rock that "had not traveled far," and geyserites and flint stones that were, "very smooth," or had traveled much further from their source.[40] Samples of these stones and more waters were taken for further analysis.

But Voigtel and Schneider mentioned another common use for mineral fragments of this sort, providing us with further insight into the theory and practice of dowsing. According to the manuals we will review in Chapter Three, when a dowser held newly exposed mineral chunks beside the dowsing rod, the attractive virtue of the rod markedly increased. The experienced dowser, in other words, could amplify

---

[39] BAF, Loc. 40013, Nr. 174, Actum, 6b.

[40] "…etliche Geschiebe in Sande gefunden worden, so den ansehen nach nicht weit gegangen Schwefelicht und Alaunisch seyn mögen, die dabey befindl. Kiesel- und Feyersteine aber waren sehr glatt gewaschen, musten also wohl weiter alß jene geführet worden seyn." BAF, Loc. 40013, Nr. 174, Actum, 6b-7. Here as elsewhere, we cannot be certain what minerals our actors discovered. 'Kieselstein' was likely geyserit, found near mineral springs, but other possibilities include chrysokoll, cerit, chalcedon, and kieserite.

the attraction a rod had for a given mineral earth below by imparting the virtue of that mineral to the rod itself, activating the rod, as it were, with a fresh mineral fragment. The logic behind this practice was related to the more common experience of magnetizing an iron needle by bringing it into contact with magnetite, or lodestone. After touching the needle with magnetite the needle points northward, as all sailors and surveyors working with compasses knew. The extention into dowsing was certainly contested: a well-known author in Leipzig, Johann Praetorius (1630–1680), had defended dowsing by making this comparison to the compass;[41] the famous Jesuit scholar, Athanasius Kircher, conducted experiments with tree branches to disprove the analogy;[42] and Johann Sperling at Wittenberg was equally intent on undermining it (Chapter Two). Göbels was the experienced dowser in this case, and the officials promptly requested that he proceed with mineral fragments in hand, both in the vicinity of the dig and at another nearby swamp, notwithstanding Göbels's declining physical condition. The results were not encouraging: the rod did not detect anything that it had not already indicated.

Voigtel and Schneider then repeated their request that the dowser not concentrate on salt alone, but rather focus "impressions and thoughts" on all mineral waters, even if they had no effect on the rod. The assumption here was, again, that the mind of the dowser played a role in how the instrument functioned; that the rod's natural attraction also required proper control or mediation by the practitioner.[43] The party assumed some sort of natural affinity between the mineral fragment and substances underground, but they allowed that the dowser's concentration and imagination was another factor in how the dowsing rod dipped. It may be that only the officials held this belief about a dowser's 'Impression,' and that the practitioners maintained only that the rod had more physical affinity with underground formations, but we do not wish to stretch the evidence.

---

[41] Johann Praetorius, *Gazophylaci Gaudium, das ist Ein Ausbund von Wündschel-Ruthen oder sehr lustreiche, und ergetzliche Historien von wunderseltzamen Erfindungen der Schätze, so geschehen seynd* (Leipzig, 1667).

[42] Athanasius Kircher, *Magnes sive de arte magnetica* (Rome, 1641).

[43] "Mann ließe aber auch dem Ruthen-Gänger Göbeln… dergl. geschiebe in die Hand nehmen, und in der Gegend mit der Ruthen gehen; alleine solche wolte darauff nirgend ziehen, sondern schlüge nur auff die ein mahl angegebene Saltz-Ader. Hierüber errinnerte mann ihm, daß er, ohne auf Saltz alleine seine Impression und gedancken zurichten, nur schlechterdings auf Sohle die Ruthe führen sollte." BAF, Loc. 40013, Nr. 174, Actum, 7.

In any case, Göbels's rod, after he had redirected his mind as his superiors requested, began to respond to new waters further down the stream. He indicated a weak salt source beside a small swamp ('h' at the top of Fig. 5), and two other similar rivulets nearby, all of which seemed to flow toward the pond (at left) into which the stream dumped. The authors reported these statements, but they also added that their dowser's weak constitution might have adversely affected the accuracy of his practice.[44] It is not clear whether the dowsing itself contributed to Göbels's worsening condition (as we have seen elsewhere), but this is another indication that the clarity and quality of a practitioner's mind—or so the officials believed—had great impact on his practice.

The party wished to discover additional mineral fragments, but water management, Göbels's sickness, and labor costs all converged to complicate matters. The wooden conduits continued to redirect waters, but even the sands upholding the pipe-works threatened to wash away, and the decision was made to abandon the dig until the workers could apply "greater force," by which the authors meant a water-raising apparatus and more advanced wooden walls and supports.[45]

In fact, the men proposed an overshot water wheel back near the old spring (illustrated by the crescent moon symbol on Fig. 2). The officials, frustrated by the ongoing battle against the mountain waters but confident in their progress here, decided to lay conduits to feed a proposed water wheel. Waters from the spring above and other areas would be channeled downward to turn a wheel, hitting it from above (undershot wheels were more common beside rivers, as for mills). Schreiber has sketched both the proposed channels between the spring and dig (dotted lines on Fig. 2), and the wheel itself (just below the crescent moon symbol). The wheel could be used to raise buckets of mountain water, and hopefully deeper-flowing brine as they became more distinct. City council representative Francke and the head forester to Pratau came to study the site and plans with Voigtel and Schneider. The hope was that such an investment would move water "with fewer costs" than at present.[46]

Costs mounted further after an unfortunate accident the digger named Richter suffered on July 9. Just before completing his 12-hour shift, as he was chopping wood to supply the Spitzberg dig, Richter

[44] BAF, Loc. 40013, Nr. 174, Actum, 7b.
[45] BAF, Loc. 40013, Nr. 174, Actum, 8.
[46] BAF, Loc. 40013, Nr. 174, Actum 9b.

accidentally struck his foot with an axe. The bleeding would not stop, and he fell unconscious. Voigtel and Schneider bound Richter's leg to help stop the bleeding, and carried him back to Göbels's place on horseback. They sent for the hangman of Gräfenhainichen, a town just to the west. The early-modern hangman, we must recall, functioned not only to separate limbs on the scaffold, for which he was also known, but as a popular surgeon and healer.[47] This is the last we hear of poor Richter, since the officials immediately placed a new digger in the foreman's care, and hired two woodcutters, perhaps aware now of the dangers of overtaxing the diggers. On July 16, the officials would hire two more diggers from Wittenberg.

One of the new woodcutters shared a third tale of salt on the Spitzberg. Martin Kießler "had boasted" (*sich gerühmt*) that he knew a nearby salt spring, and he proposed to lead the way. The officials called on none other than Göbels to confirm the latest claim, but his condition finally compelled him to decline the request. He suggested that the peasant dowser, Fischer, come in his stead, but the authors recalled their poor estimation of Fischer, opted to release Göbels, and did not entertain Kießler's claim.[48] This is the last we hear of Göbels and his 'house.'

The final entries in the 'Actum, Göbels Haus' indicate that newly hired workers were making progress laying conduits, but that waters were taking the upper hand once again. Two new workers had put down some 36 Lachter (72 meters) of channels by July 19, but council representative Francke was shown the latest water encroachment and precarious conditions at the dig itself.[49] Additional letters reveal a continued expectation that Spitzberg would produce salt, but we might end our story here. Voigtel and Schneider wrote to Dresden on July 21, conceding that Fischer's original story had proven false, but emphasizing that exploration had proceeded nonetheless, that the Freiberg dowsers had indicated promising spots, and that with better equipment, especially a water wheel, the region could become productive. Privy Counselor Alemann asked for a formal opinion (*Gutachten*) from Bergrat Littman in Dresden, and Alemann's own letter of September 22, 1714, confirms that work would go on. A letter dated February 1715 mentions poor weather.

[47] BAF, Loc. 40013, Nr. 174, Actum, 8–8b.
[48] BAF, Loc. 40013, Nr. 174, Actum, 8b.
[49] BAF, Loc. 40013, Nr. 174, Actum, 10.

And so it must have continued, month after month, winter thaw after winter thaw, in search of the white gold that was making Prussia wealthy, but that continued to elude Saxony. The fate of Spitzberg Hill may be recorded elsewhere in additional bundles archived by the Saxon mining jurisdictions, but hindsight tells us that no major salt works appeared here. The Elector had greater success with Borlach at the nearby towns of Artern, Kösen, and Dürrenberg.

Mining Science

In a textbook on mine surveying of 1782, the mathematician, Johann Friedrich Lempe, claimed that prior works in this field were too "craft oriented" (*handwerksmäßig*). Lempe wished rather to present mine surveying as a veritable science, and he cited the 1749 text of the joint-founder of the Freiberg Mining Academy, Christian Friedrich Wilhelm von Oppel, as a worthy model.[50] The mine surveying we witnessed in this chapter was certainly not the science that Lempe meant, but a more craft-oriented practice in which mining officials worked closely with dowsers and diggers in the field, and entertained the knowledge of peasants like Fischer. Mathematical surveying and dowsing were not incompatible activities for Voigtel, but complementary practices, as the surveyor confronted rough conditions in the field, and relied on those with a greater 'feel' for the mountain than he alone possessed. When *Bergwissenschaft* was a predominately vernacular science, such a direct line between local and professional knowledge was still possible. We shall return to mining books such as Voigtel's in Chapter Three. But first, how did the rod become the legitimate instrument of the proud miner? How did it escape charges of sorcery and witchcraft to number among the mine surveyor's tools? Why did it persist in practice when mining spirits and other beliefs declined?

---

[50] Johann Friedrich Lempe, *Gründliche Anleitung zur Markscheidekunst* (Leipzig, 1782), 9. Lempe studied mathematics at the Freiberg Mining Academy and, following publication of this text, was called to Freiberg to instruct in mathematics and mining machines. ADB, s.v., 'Lempe, Johann Friedrich.'

CHAPTER TWO

MAGIC, WITCHCRAFT, AND THE NATURE OF THE ROD

> One day, at Goldenhöhe, a miner was carrying a chunk of mineral ore when he saw a mining gnome. Since the miner forgot to give the proper greeting, the spirit grabbed the ore and smashed it to the floor into a thousand pieces. Twisting and screaming, it disappeared, and the earth trembled. When the miner recovered from his shock, he told his story, and never forgot to greet the *Berggeist* again.[1]

> Now and again there were spirits in the mountains, especially at the mines: small gray creatures wearing large felt hats. They abducted and killed miners when they did not behave properly. They were to stand to the right and keep silent.[2]

These were the spirits of the Erzgebirge, born of the waving shadows, eerie sounds, and grotesque forms that confounded the underground worker. Deep in the bowels of the earth, the real miner felt as small as these diminutive diggers in miner garb. The cause of countless accidents and misfortunes underground, the spirits could also warn of impending dangers and unsafe conditions. There were good and evil mountain spirits. So deeply rooted were they in mining culture, in 1869, the folklorist Wenzel Peiter could warn against challenging the "old miner" of the Erzgebirge about his belief in *Berggeister*.[3]

But talk of mining spirits had in fact declined markedly since the Reformation. Dogmatic clerics, scholars, and state officials preoccupied with religious orthodoxy and social control had sought to reform mining culture. Scholars of the early sixteenth century, such as Paracelsus and Georg Agricola, acknowledged both good and evil spirits, but authors writing some decades later denounced all varieties. Jean Bodin, a major Catholic theorist of witchcraft, identified even the helpful mining spirit with Satan in his influential demonology of 1580, *De la Demonomanie des Sorciers*. A Protestant preacher in the mining town of Scheibenberg, Saxony, Christian Lehmann (1611–1688),

---

[1] Gerhard Heilfurth, *Bergbau und Bergmann in der deutschsprachigen Sagenüberlieferung Mitteleuropas* (Marburg: N.G. Elwert Verlag, 1967), 535.
[2] Heilfurth, *Bergbau und Bergmann*, 537.
[3] In Heilfurth, *Bergbau und Bergmann*, 177.

reduced all sightings to diabolical delusions,[4] and Caspar Posner published an important essay on spirits at Jena in 1662. He reviewed learned authorities from the days of Agricola, but considered the Devil to be the author of all sightings.[5] According to the folklorist, Fritz Roth, the more generic name 'devil' gradually replaced individual and local names for mining spirits.[6] Enlightenment-era social reformers and educators largely completed the eradication of belief in mining spirits among the lower orders that earlier theologians had begun. Today, one is hard-pressed to meet anybody in the Erzgebirge who truly believes that mining gnomes roam the hills and tunnels.

Like underground spirits, the dowsing or divining rod (*Wünschelrute*) was a major component of pre-modern mining culture. The forked stick was used in all manner of magical treasure hunting, sometimes involving mining spirits and other creatures on the mountain. Dowsing rods, like special mirrors and crystal balls, functioned as instruments of sorcery that offered hidden knowledge of the underworld. The mining folklore is rich in examples. According to one tale, a group of miners once set out with a rod to a region reputedly rich in mineral deposits. The rod dipped strongly before a loud noise echoed through the forest, and the lantern extinguished, "as if the evil spirit would appear." The miners ran home and never tried the dowsing rod again.[7] The same forces of social disciplining and confessionalization that purged the mountains of mining spirits threatened the dowser as well, but his practice survived the early-modern period more firmly identified as the distinctive instrument of the proud miner.

This chapter introduces the dowsing rod and begins charting its unique historical path. Early opponents wished to identify the practice with witchcraft. Theologians and philosophers intent on ridding the mines of diabolical influence said that practitioners made an 'implicit' pact with the Devil. However suggestive, the opinion was impotent against the widespread use of rods in mining, where a powerful theory

---

[4] "Während die Volkssage gute und böse Züge im Bilde des Bergmönches kennt, erscheint bei Lehmann nur das Verderbliche seines Wesens." Fritz Roth, *Christian Lehmanns Leben und Werke und seine Stellung zum Aberglauben* (Marburg: C.M. Gärtner, 1933), 64.

[5] "Hic enim commodiorem occasionem sumit insidias & ludos faciendi hominibus Satanas," in *Diatribe physica de virunculis metallicis*. In Heilfurth, *Bergbau und Bergmann*, 132.

[6] Roth, *Christian Lehmanns Leben*, 64.

[7] Heilfurth, *Bergbau und Bergmann*, 962.

arose to legitimize the practice. The rod responded to mineral vapors from the earth, and certain experienced and pious men—the *Bergverständige*, or 'knowers of mining'—had the God-given ability to detect these vapors. This idea echoed through mining books of the seventeenth and eighteenth century, distinguishing the practice from magic, witchcraft, and superstition, and providing an intellectual ground stone upon which new theories could build. Historians such as Robert Scribner have shown that confessionalization and social disciplining did not eliminate all magical belief, and others have recognized that the mountains often marked the limits of popular reform. "It is not easy to find Protestant areas which resisted the reform of popular culture until after 1650," confirmed Burke, "but they did exist, particularly in the mountains."[8]

The Wünschelrute

We can introduce the practice more formally, setting it within broad agricultural, domestic, and mining contexts. The dowsing rod has ancient origins, although whether ancient miners wielded the rod remains an open question. Moses produced water by striking a rock with a rod; Minerva used a rod to make Ulysses young again; and, according to some accounts, Hermes procured his famous staff from a tree—the distinctive snakes of the caduceus were a later development.[9] The Greek words *rhabdos* (rod) and *manteia* (divining) combined as *rhabdomancy*; the Romans had their *virgula divina*; and the Scythians, Persians, and Medes all used branches in divination and sorcery.

[8] Burke, *Popular Culture*, 238. Robert Scribner, "The Reformation, Popular Magic, and the 'Disenchantment of the World,'" *Journal of Interdisciplinary History* XXIII:3 (Winter, 1993). See also Dym, "Mineral Fumes and Mining Spirits"; Ronald Po-Chia Hsia, *Social Discipline in the Reformation: Central Europe 1550–1750* (New York: Routledge, 1984), 176; Kevin C. Robbins, 'Magical Emasculation, Popular Anticlericalism, and the Limits of the Reformation in Western France Circa 1590,' *Journal of Social History* (Fall, 1997), pp. 61–83; C. Scott Dixon, *The Reformation and rural society: The parishes of Brandenburg-Ansbach-Kulmbach, 1528–1603* (New York: Cambridge University Press, 1996), pp. 176–93; Susan Karant-Nunn, *Luther's Pastors: The Reformation in the Ernestine Countryside* (Philadelphia: American Philosophical Society, 1979); and Lorna Jane Abray, 'The Laity's Religion: Lutheranism in Sixteenth-Century Strasbourg,' in Ronald Po-Hsia (ed.), *The German People and the Reformation* (Ithaca: Cornell University Press, 1988), 216–232.

[9] Ludwig Preller, "Der Hermesstab," in *Ausgewählte Aufsätze aus dem Gebiete der classischen Alterthumswissenschaft* (Berlin: Weidmann, 1864).

Herodotus and Cicero mentioned the use of dowsing rods in sorcery, though Pliny discussed how to find water and minerals without mentioning them, evidence for some folklorists of its later origin, or at least of its later application in water and mineral prospecting.[10] The physicist and last major scholar of dowsing, William Barrett, argued that the more distinctively modern rod used toward these ends originated in the mining districts of Germany in the late fifteenth century, and spread from there to the rest of Europe and beyond.[11]

The dowsing rod had diverse healing and magical properties, suggesting that farmers and miners did not readily identify it with harmful magic or witchcraft. This was true of countless twigs, leaves, barks, nuts, and flowers that men and women used daily. Bächtold-Stäubli's *Dictionary of German Superstitions*, a large collection of folklore based on sources dating from the fifteenth through nineteenth centuries, contains entries for 'tree,' 'branch,' 'hazel,' 'dowsing rod,' 'treasure,' and 'treasure hunting,' that we can compare to other sources.[12] We glean some sense of the popular importance of the dowsing rod by looking at the entry for hazel, out of which the rod was frequently made. The bush was widespread in Europe, and, according to Bächtold-Stäubli, possibly the most common used in popular culture.[13] Aside from scattered instances of harmful magic, Europeans had attributed healing properties to the nuts, flowers, juices, and branches of the hazel bush since antiquity. The pervasiveness of hazel across the continent, and its early blooms, edible nuts, and flexible branches, all suggested beneficial magical qualities. For example, a hazel branch helped the lost traveler return home, and it forewarned him of insurmountable walls and dangerous drops. The soldier who carried with him seven 2-inch long hazel twigs broken off on St. John's Day (June 24th) between 11:00 and midnight was protected from harm. Vegetable matter like roots and herbs collected on June 24th, a feast day, had magical and healing properties in popular culture.[14]

---

[10] Evon Z. Vogt and Peggy Golde, "Some Aspects of the Folklore of Water Witching in the United States," *Journal of American Folklore* 71:282 (1958), 522.

[11] Barrett and Besterman, *The Divining Rod*.

[12] Bächtold-Stäubli, *Handwörterbuch des deutschen Aberglaubens* (Berlin and Leipzig: W. de Gruyter & Co., 1927–1942), s.v. "Bergbau," "Berggeist," "Witterung," "Wünschelrute," "Schatz," "Schatzgraben."

[13] "...darf wohl...als der volkstümlichste Strauch bezeichnet werden." Bächtold-Stäubli, *Handwörterbuch*, s.v. "Hasel."

[14] Karl Wegert, *Popular Culture, Crime, and Social Control in 18th-Century Württemberg* (Stuttgart: Franz Steiner Verlag Wiesbaden, 1994), 52.

In the domestic sphere, likely in the hands of women, hazel cast unwanted bugs and animals from the home, and it scared off snakes. The hazel nut helped milk turn to butter. Hazel was ground into powder and used in love potions. It was rubbed on warts and over broken limbs, and it was sprinkled into open wounds to expedite healing.[15] Other popular uses included warding off evil spirits and counteracting witchcraft, as when hazel twigs were embedded in manure on Walpurgis Night (April 30). Farmers beat their bewitched livestock with hazel or fed them the flowers or twigs. To see witches, one might carry a hazel branch, properly acquired, into the fields at sunrise. To see the elves, take a three-pronged hazel branch between August 15 and September 8 (*Frauendreissiger*), cut off the middle branch, and lay it under a bed.

The dowsing rod was also three-pronged, which suggests that it conformed to a standard form. The folklore suggests that the rod had countless uses, and that it was well integrated into popular magic. Bächtold-Stäubli includes an entry for '*Wünschelrute*' of some eighteen pages, noting that the majority of tales come from German mining regions, where it was also known as '*Visierrute*,' '*Schlagrute*,' '*Fragerute*,' '*Wickerode*,' and '*Wickerraue*.' In France, it was the '*Baton de Jacob*' or '*Baguette divinatoire*,' in Holland the '*Toverstaf*' or '*Wichelstuk*,' and in England and America, the 'divining' or 'dowsing' rod, but also 'witch stick,' as in 'water-witching.' The branch may have earned this title because many were made of witch hazel (which evidently had a prior association with witchcraft). The rod was also commonly used in Switzerland, Scotland, and Scandinavia.[16]

An anonymous manual dated 1668 explained how the rod worked both in mining and everyday life, providing step-by-step instructions.[17] The dowsing rod could be made of beech, fir, ash, oak, willow, and apple- or pear tree. Iron or brass wires and stakes fashioned into dowsing rods were also possible, though less common before the eighteenth century. Bächtold-Stäubli distinguishes between the 'living' (wooden) and 'dead' (metallic) varieties of rod. Animal bones and horseshoes were other dead varieties. Most practitioners used the hazel branch, a

[15] Bächtold-Stäubli, *Handwörterbuch*, s.v. "Hasel."

[16] Vogt and Golde, "Some Aspects of the Folklore of Water Witching." Theodore Besterman, "The Folkore of Dowsing," *Transactions of the Folklore Society* XXXVII:2 (1926).

[17] Sächsisches Hauptstaatsarchiv Dresden [DHSA], Loc. 36062.

testament to its role in popular culture as a symbol of fertility, but also related to the widely recognized "fact" that it was "hardly ever hit by lightning."[18] As a first step, the anonymous manual similarly recommended that the dowser break from "whichever wood he wants," and that when none were available, that he use a metal wire.[19] 'Breaking' the rod from its tree or bush was a popular idiom. A one-year-old, unscarred, three-pronged hazel branch was the prototypical dowsing rod, though the exact shape and length varied considerably: between roughly one and five feet long, according to Bächtold-Stäubli.

Popular wisdom on the opportune time and manner in which to break the branch was specific, but it varied from region to region, and from practitioner to practitioner.[20] The user might break the rod on Christmas Eve, New Year's Eve, Good Friday, Walpurgis Night, Saint John's Day, or Ascension Day, during Matins, and during a full moon or the first Sunday of a new moon, and especially at midnight. The most common time was at midnight on St. John's Day. The user chose a one-year-old branch located on the bush so that the midday sun shone directly through the two shorter legs, and it received both the morning and evening sun. The prospective dowser approached the tree silently. Prayers, spells, or invocations were recited at the moment of breaking, frequently, "in the name of the Holy Trinity."[21] After breaking, the rod, like the Apostle John himself, might be baptized with holy water: "I baptize you, Saint John's Rod, in the name of God the Father, the Son, and the Holy Spirit; I baptize you, like Jesus baptized John in the River Jordan...That you show me where something lays buried or walled up... that you dip forward where something is, and stand still where nothing is...as true as Mary sacrificed Jesus at the altar [the Purification]."[22]

Christian doctrine and practice combined quite nicely with popular beliefs, though the mostly Protestant authors in Saxony did away with ceremonies surrounding the rod, and appealed to other explanations. The 1668 manual may expose Protestant authorship by denouncing appeal to the stars, magical formula, and other such acquisition rituals

---

[18] Bächtold-Stäubli, *Handwörterbuch*, s.v. "Wünschelrute."
[19] DHSA, Loc. 36062, Fol. 6b.
[20] Bächtold-Stäubli, *Handwörterbuch*, s.v. "Wünschelrute."
[21] Bächtold-Stäubli, *Handwörterbuch*, s.v. "Wünschelrute."
[22] Bächtold-Stäubli, *Handwörterbuch*, s.v. "Wünschelrute."

as "superstition."[23] It did not explain the mechanism behind the dipping motion in any detail beyond tying it squarely to the spiritual condition of the dowser: "God did not attribute the power to the rod, but rather to man."[24] After the Fall, mankind lost its natural understanding of such hidden mysteries and capabilities. Other sources claimed that only a respectable and pious man could break the rod, or, alternatively, an innocent child, born on Sunday, and raised in a properly Christian home.[25] The practitioners we shall discuss also professed great piety, seeing no contradiction between simple Christian faith and dowsing practice. Dowsing required a clear conscience and an upstanding life.

Like hazel more broadly, the dowsing rod had diverse uses in agriculture, mining, and everyday life. Much depended on the idiosyncratic method of the practitioner, but standard practices emerge. The user grasped the two shorter legs in each of his hands, palms up (as if taking the reins of a horse), between the middle and index fingers; he then raised the rod to breast level, holding the main shaft upwards. He proceeded in silence, handling the rod loosely, and making an effort to rid his mind of preconceived expectations and other distracting thoughts. When the rod detected something below, it dipped downwards, revolved, or otherwise shook, movements that sometimes extended to the body of the dowser. If held too tightly, the rod might break.[26] It responded to lost objects, underground springs, boundary markings, lost livestock, and even the perpetrator of a crime in 1692.[27] It also divined secret knowledge, such as the fate of the soul after death, the health of distant relatives, the sex of the unborn, and even the truth or falsity of information.

The manual was mostly concerned with small-scale mining, but it also included instructions for more everyday applications. To determine the veracity of statements or acquire knowledge of future events, it recommended preparing a table with 'Yes' 'No' and 'Maybe' columns,

---

[23] "Dass wegen ist in gegen satz alles vergebl: dass etl: sich an gewisse Tage, Stünden und andern Caracteres binden solche zu brechen, damit solche durch aberglauben bey Bergkwerken was thun sollen, denn solches ist ein Aberglaub, so wieder dass erste Geboth." DHSA, Loc. 36062, Fol. 4b.

[24] "Gott hat nicht die Krafft dem Holtzen sondern dem Menschen zu geeignet." DHSA, Loc. 36062, Fol. 5.

[25] Bächtold-Stäubli, *Handwörterbuch*, s.v. "Wünschelrute."

[26] Bächtold-Stäubli, *Handwörterbuch*, s.v. "Wünschelrute."

[27] See Chapter Five.

to which the rod would dip, depending on the question.[28] Books on how to interpret patterns, lines, lots, and figures (geomancy) filled with questions and tables such as this were common. One example in the Saxon Government Library includes entries on whether something would increase or decrease, what fruit one should harvest, which part of the body was ailing, whether something was far or nearby, and such mining-related inquiries as, "Is the vein for which you are digging overhanging or underlying?" "Should the workers dig a gallery or shaft?" And even, "Is the shaft manager or foreman a knave?"[29]

Another form of dowsing magic involved a detailed chart. The manual instructed the dowser to sketch a compass face on a piece of paper and plot the names of territories and mines, as well as the astrological signs of the known minerals. The rod would actually dip to the chart at the most promising spot in the field, much like a modern-day ouija board.[30] Mining charts like this, surveyed by the rod, would also divine the depth, layout, and quality of the mineral vein. The rod might also dip to a common land map.[31] In the field, according to the 1668 tract again, the number of dips indicated the sort of ore: three times for mercury, six times for bismuth.[32] The number increased for alum (18), sulfur (9), cobalt or gold (28), or iron (30), according to no obvious logic or reason.[33]

There were as many explanations for the strange motions of the rod as there were dowsers. Some believed that the rod dipped only to simpletons or those graced by God. Practitioners in mining tended to agree, but they also assumed a more natural correspondence between metal and rod, and smeared it with substances like marcasite

---

[28] DHSA, Loc. 36062, Fol. 17b.

[29] *Das erste Buch Nein*, Sächsische Landesbibliothek, Staats- und Universitätsbibliothek Dresden [SLUB], Sondersammlung, Schrank N, Nr. 7, Bl. 188, 186, 235.

[30] "Vors andern...lege er die vorgeschriebenen Tabell...vor sich, und Lassen Ihn einen Bergk oder eine ganzen Landschafft benahmen schreibe dann unter bemeldte Tabell den Nahmen der Landschafft oder des Berges und lass den Ruthen auf die verzeichneten planeten u. Charact: schlag, auf welche sie nun schläget, dasselbigen Metall od. Mineral ist in der darunder verzeichneten Landschafft od. Berg vorhanden." DHSA, Loc. 36062, Fol. 6b.

[31] Bächtold-Stäubli, *Handwörterbuch*, s.v. "Wünschelrute."

[32] "Sechstens mustu wißen, was vor Metall oder Mineral wo dir die ruthe geschlagen...in gängen verborgen, solches wirstu folgender Maßen...zu finden er kundig, als schlägt dir die ruthe 3 mahl, so habe ich Quecksilber, item 6. mahl Wißmuth 9. mal...." DHSA, Loc. 36062, Fol. 8.

[33] DHSA, Loc. 36062, Fol. 9b.

(something akin to iron pyrite), or fixed metal into the rod to increase its attractive virtue. They compared this attraction to that between iron and a lodestone, mercury and gold (as in the process of amalgamation), or the sun and plant life. Other theories emerged among scholars and mining officials, involving correspondence between spatially separated objects, occult qualities, mineral vapors, magnetism, and after the eighteenth century, electricity, galvanism, and even psychological suggestion. Those who rejected dowsing likened the practice to casting lots, gambling, sorcery, witchcraft, and other crimes described in the law ordinances. The practice was most suspect when users claimed that God, the angels, or spirits of uncertain origin were the immediate causes of the motion. The 1668 manual, which blurred the distinction between geomancy and Christian piety, demonstrates just how difficult it was to make these distinctions.

Dowsing manuals also contained rod invocations (*Rutenbeschwörungen*) intended to secure or 'bind' supernatural powers. Bächtold-Stäubli suggests that such formulas developed in the fifteenth century as popular versions of learned, Latin prayers of much earlier date.[34] One example comes from an anonymous eighteenth-century collection of incantations to bind spirits into performing a number of everyday functions, including giving protection against misfortune, securing the love of another, and finding lost treasure. This work explained how to break and bind a branch for magical treasure hunting, appealing to both pagan spirits, and Old and New Testament powers:

> In the high and holy names of God and the Divine Majesty *Adomaton, Amaray, Amath, Honelle, Jahweh, Eleison*...I break you from your trunk, creature and created wood, so that you have the power and virtue to point and show me, [Name], all hidden treasures, be it silver or gold, in the earth or other hidden places. So that you show me where it is, through the sign of dipping down; where it is, whether the treasure is behind you, before you, or beneath you. Give it this sign and dip downwards strongly. Do this rod, in the sacred name of the Holy of Holies and Highest Glory and One Trinity Yahweh. Amen. Amen. Amen.[35]

---

[34] A fourteenth-century herbal contained this prayer for the 'binding' of a root: 'In nominee patris quero te, i.n. filii invenio te, et i.n. sp. st. te carpo, ut sis mihi et omnibus...obstaculum contra omnia seva jacula omnium inimicorum nostrorum.' Bächtold-Stäubli, *Handwörterbuch*, s.v. "Wünschelrute."

[35] *De virgula divina*, Gotha Landes- und Forschungsbibliothek [GFB], Ch.B.1432, Bl. 2.

The same text also had a spell for investing a certain shield with the power of the Holy Trinity to drive away evil guardian spirits from discovered treasure.[36]

Anonymous collections of incantations, conjurations, and instructions on fashioning instruments of sorcery became a popular genre in the late sixteenth century, a period of fascination with Satan and the Faust legend, when, says William Clark, "the Devil had most success" in entering the popular imagination.[37] They bore titles such as "True Art of Hell-Binding and Treasure Hunting,"[38] "The Hell Binding [Art] of Faust"[39] "The Mirror of Solomon,"[40] and "A Book on How to Capture the Spirits,"[41] and they were used to discover mineral ore, among other treasures. Eva Labouvie discusses a number of manuals from the early nineteenth century. In one, a crystal-ball manual, the incantation reads: "I [Name] purify you from…all evil spirits…and give you this name…[so] that you lead to the true and proper spot of hidden goods, like hidden treasure of gold and silver…and all mineral veins underground and in mining works…in the name of God the Father, God the Son, and God the Holy Spirit. Amen."[42] The formula was essentially unchanged from the sixteenth century.

These popular manuals presented the dowsing rod as one among many instruments that could be used for magical treasure hunting, such as crystal balls, magical mirrors, polished stones, water pools, enchanted swords, and even fingernails. In the mining town of Scheibenberg, Pastor Christian Lehmann (d.1688) knew of a monk in

[36] "O Aller heiligste und hochgelobte Dreyfaltigkeit du unergründliche und unzertennliche drey einigkeit, seegne durch deine göttliche Krafft dieser Schild, daß es alle Höllische geister fürchten, daß sie sich dafur erschrecken und fliehen mussen, als für deiner Gottlichen majestät persönlich, daß ich sie damit kan von denen verborgenen Schätzen treiben, und davor weichen und fliehen müssen und nicht eine eintzige macht davor beweisen können, die mir in meinem thun und an meinem Leibe und an mein erseelen und an meinem Leben schädlich wäre. Dieser bitte ich von dir… durch Jesum Christum. Amen Amen Amen." GFB, Ch.B. 1432, Bl. 16.

[37] William Clark, "Der Untergang der Astrologie in der deutschen Barockzeit," in *Im Zeichen der Krise: Religiosität im Europa des 17. Jahrhunderts*, Harmut Lehmann and Anne-Charlott Trepp, eds. (Göttingen: Vandenhoeck & Ruprecht, 1999), 442.

[38] *Der rechte Geistliche Gezwang der Höllen und Schatzgräber Kunst*, SLUB, Schrank N, Nr. 87.

[39] *Fausts Höllenzwang*, DHSA, Loc. 4417, Fol. 12. *Johannis Faustus Höllen-Zwang*, SLUB, Schrank N, Nr. 165b.

[40] *Speculum Salomonis*, SLUB, Schrank N, Nr. 166.

[41] *Ein Buch zumachen da Geister innen verschlossen*, SLUB, Schrank N, Nr. 166.

[42] Labouvie, *Verbotene Künste*, 115.

Kamenz who possessed crystal balls, magic seals (*Geistersiegel*), and dowsing rods, among other instruments.[43] Participating in a new wave of clerical critique highly sensitive to witchcraft, Lehmann was deeply hostile to these activities, warning against the "suspicious books" that gave instruction in these methods.[44] One such book, purporting to teach how to "conjure and invoke the spirits," included a rather striking image of the Evil One himself, sporting a robe and sandals, and carrying a bell and scroll.[45] The author, who identifies himself as Johann Faust, wished to share the mysteries of magical treasure hunting. One must mind the spirits guarding the presumed treasure, he says, watch for candles that extinguish suddenly, and proceed only on astrologically propitious days. When all proper care is taken, the dowsing rod is sure to reveal the treasure, if not a mineral vein crossing below.[46] Above all, the treasure hunter must trust in God, since the Devil can tempt the unsuspecting hunter into danger: he might, "make the rod, crystal ball, or mirror speak truly one or more times."[47] The author proceeded with instructions on how and when to fashion the requisite candle to be used with the dowsing rod.

Dowsing and Witchcraft

That Dr. Faust himself may have wielded the rod suggests that dowsers engaged in black magic. But did secular or religious authorities accuse practitioners of making a pact with the Devil? Were dowsers prosecuted like witches? Luther had formally rejected astrology and dowsing as early as 1518 in an extended commentary on the Ten Commandments. Among the practices banned by the First Commandment, Luther listed crystal ball gazing, healing with special ointments, sorcery with magical circles, and most everything offered by so-called

---

[43] Roth, *Christian Lehmanns Leben*, 84. See also Stefan Jäggi, "Alaunenhändler, Schatzgräber und Schatzbeter im alten Staat Luzern des 16.–18. Jahrhunderts," in *Mitteilungendes Historischen Vereins der fünf Orte Luzern, Uri, Schwyz, Unterwalden ob und nid dem Wald un Zug* 146 (1993).

[44] "Hütet euch für den verdächtigen Buch eures Mitbürgers. Denn der Satan ist tausendlistig und verdeckt seine Zauberkunst unter solchen astralischen Characteribus, Planetenfiguren, Amuletis, Divinationius u.a. curiosen Narrentheidungen...ins Feuer mit solchen Schrifften." Roth, *Christian Lehmanns Leben*, 85.

[45] DHSA, Loc. 4417, Nr. 13, Fol. 1.

[46] DHSA, Loc. 4417, Nr. 13, Fol. 5b.

[47] DHSA, Loc. 4417, Nr. 13, Fol. 5b.

wise men and women. "Those who search for hidden treasure with a dowsing rod" were equally guilty of worshipping a false idol.[48] His follower, Johann Spangenberg, wrote that children should be instructed that the Devil was the treasure-hunter's true master.[49] However influential Luther's opinion was, his famous contemporaries at Wittenberg, Melanchton and Peucer, both accepted astrology and a theory of sympathy that would have allowed for dowsing.[50] There was no official Wittenberg position on the rod during the Reformation.

This would change during the witch-hunts, which encouraged a more thoroughgoing condemnation of dowsing. Witchcraft persecutions peaked in Saxony between 1600 and 1630, and again between 1660 and 1680.[51] Rolf Schulte shows that an average of 24% of all witchcraft accusations in Germany were directed toward men, a higher number than the 20% often cited in the historiography.[52] The percentage of executions among accused men, however, was lower than that among accused women. Wilde discovers that men made up 22.9% of all known accusations in Saxony, where the sex of the defendant is known, and only 13.4% of the executions.[53]

It was in this context that the Wittenberg professor of philosophy, Johann Sperling (1603–58), developed a correlation between witchcraft and dowsing that defenders would have to confront.[54] Wittenberg scholars proved especially interested in dowsing, mining spirits, and

[48] "Item Qui occulta quaerunt per Christallum, unguem, vel eberneum capulum, tamen prius consecratu. Nec id possunt, nisi puellae vel pueri virgins [sic], in circulo inclusi, ne desit species bona pessimo illi operi. Dicitur tamen huius inspectionis mysterium fallacissimum esse. Huius generis sunt. Qui virga divinationis occultos quaerunt thezauros. Maxime autem hi, quos vulgo, viros & mulieres sapientes (i. Magos ex phitonissas) vocant, Quoru [sic] oracula & diversis et longinquis locis petut [sic]." Martin Luther, *Decem Praecepta Wittenbergesni praedicata populo* (Leipzig, 1521).

[49] Stuart Clark, *Thinking with Demons: The Idea of Witchcraft in Early Modern Europe* (New York: Oxford University Press, 1999), 491.

[50] Kaspar Peucer, *Commentarius de praecipuis divinationum generibus* (1553). See Sachiko Kusukawa, "*Aspectio divinorum operum*: Melanchton and astrology for Lutheran medics," in Ole Peter Grell and Andrew Cunningham, eds., *Medicine and the Reformation* (New York: Routledge, 1993), 33–56.

[51] Manfred Wilde, *Die Zauberei- und Hexenprozesse in Kursachsen* (Cologne and Weimar: Böhlau Verlag, 2003), 158.

[52] "Die Zahlen verweisen auf eine relative Heterogenität in der Verfolgungspraxis, denn jede vierte Hexe war ein Mann." Rolf Schulte, *Hexenmeister: Die Verfolgung von Männern im Rahmen der Hexenverfolgung von 1530–1730 im Alten Reich* (Frankfurt am Main: Peter Lang, 2001), 81.

[53] Wilde, *Die Zauberei- und Hexenprozesse*, 308. See also Lara Apps, *Male Witches in Early Modern Europe* (Manchester, NY: Manchester University Press, 2003).

[54] See *Allgemeine Deutsche Biographie*, s.v., 'Sperling, Joh. S."

other mining-related matters.[55] In *Institutiones Physicae* of 1649, Sperling identified Satan as the causative agent behind mining spirits. Sperling's posthumous piece on dowsing, the *Disquisitio Philosophica*, which concluded that the rod functioned by implicit pact with the Devil (*ex pacto implicito*), incited a local tradition of commentary, and provided critical-minded officials, clerics, and scholars into the eighteenth century with a powerful set of arguments. Most of these were drawn from another work, the *Synopsis Physica* of 1645 (dedicated to a pastor and superintendent at Freiberg), in which Sperling treated all aspects of natural philosophy.

Sperling's *Disquisitio* is divided into three parts: the first considers the etymology of 'dowsing rod,' the second explores learned opinions on the subject, and the third is concerned with the question of whether the rod dipped via occult quality. The concept of occult qualities, which Sperling rejected in this case, requires some explanation. Natural philosophers of Sperling's time distinguished between manifest qualities (*manifestas qualitates*) and occult or hidden qualities (*occultas qualitates*), both considered real properties of nature ('occult' was distinct in meaning from its later, derogatory connotation). The difference lay in whether the cause of a given phenomenon was recognized by the senses. Hotness, wetness, hardness, roughness, brittleness and their opposites were, for example, tangible. Occult qualities included the attraction between a lodestone and iron, planetary influences, the action of a purgative or poison on the body, the spread of fire, and the shock of an electric eel. According to the Jesuit, Duarte Madeira Arrais, though occult qualities are "manifest to the intellect, they are not apparent to the senses."[56] Sometimes an object affected by an occult quality showed a visible analogy or 'sympathy' with its counterpart. Keith Hutchison argues that medieval Christian Aristotelianism mostly excluded occult phenomena such as these from knowledge, but that major seventeenth-century natural philosophers such as Daniel Sennert, William Gilbert, René Descartes, and Robert Boyle

[55] Barrett and Besterman list thirteen dissertations between 1661 and 1764, mostly from Wittenberg, in *The Dowsing Rod*. These include: Johanne Sperling, *An Virgula mercurialis agat ex Occulta Qualitate, Disquisitio Philosophica* (Wittenberg, 1666); Casparo Posnero, *Diatribe Physica de Virunculis Metallicis* (Jena, 1662); and Theodorus Kirchmayer, *Dissertatio Physica de Virgula Divinatrice* (Wittenberg, 1679).

[56] In *New Philosophy and Medicine concerning Occult Qualities* (Lisbon, 1650), quoted in J.L. Heilbron, *Electricity in the 17th and 18th Centuries: A Study of Early Modern Physics* (Berkeley, CA: University of California Press, 1979), 24.

overcame this "major epistemological impasse" by reconsidering their existence, believing now that neither occult nor manifest qualities were sensible.[57]

Sperling accepted occult qualities,[58] citing Sennert, and reviewed such examples as the lodestone and spread of fire: "Oil pulls fire to itself at a distance and is inflamed."[59] In *Synopsis Physica*, he treated the issue in greater detail. In the *Disquisitio Philosophica* on dowsing, Sperling wished merely to undermine the argument that the rod functioned by another such insensible occult quality, addressing in particular the association between the lodestone and rod, which was a suggestive parallel example. The lodestone, Sperling began, attracted iron at all times and places, with the same intensity, and without human intercession. None of this was true of the dowsing rod. Why was it that the rod worked for some, but not for many? Why did the human bearer influence the supposed occult quality?[60] If the rod operated according to an occult quality, why did it not operate when the branch remained on the tree? Why did it function only when removed and "dead?" Would not the occult quality be stronger on the living tree?[61] Perhaps some aspect of the person advanced or impeded the presumed occult quality, affecting the rod at the hands. Invoking recent philosophical theories, Sperling allowed that the hands released an "elemental fluid" (*effluvia elementorum*). He denied, however, that such a manifest quality would have any bearing on the occult.[62]

What, then, accounted for the widespread belief in the dipping motion of the dowsing rod? The answer was obvious to Sperling: the rod did not work due to some quality in nature, but rather to another cause, a major source of deception. Satan was the author of all deceit in

---

[57] Keith Hutchison, "What Happened to Occult Qualities in the Scientific Revolution," in *The Scientific Enterprise in Early Modern Europe*, Peter Dear, ed. (Chicago, 1997), 86.

[58] "Occultas qualitates non negamus...multa fiunt per manifesta, multa per occultas." Sperling, *Disquisitio Philosophica*, B3.

[59] "Naphtha et longinquo ad se trabit ignem et inflammatur." Sperling, *Disquisitio Philosophica*, C.

[60] "Magnes trahit ferrum...per occultam qualitatem, omni loco et tempore, nullo amplius interveniente medio, nullo intercedente auxilio...Et quam multi ex hominum genere sunt, quibus tam beatis esse non licet?" Sperling, *Disquisitio Philosophica*, B4.

[61] "Cur virgula non inclinat se ad metallum quando adhuc cum suo toto est, quando nondum abscissa?" Sperling, *Disquisitio Philosophica*, C.

[62] "Non negamus, egredi effluvia ex manu, uti etiam ex toto corpore, nimirum effluvia elementorum egrediuntur...at haec per manifestas qualitates agunt, non per occultas." Sperling, *Disquisitio Philosophica*, C2.

the world, and he led humankind into the most extraordinary errors.[63] Sperling had already argued in 1649 that mining spirits, both good and bad, derived from satanic influence.[64] Now, defying all those who would identify the dowsing rod as a form of harmless or natural magic, Sperling invoked a tradition that stemmed from Thomas Aquinas, and which later demonologists further developed, of distinguishing between contracts explicitly made with Satan and those accepted implicitly or less consciously. An original pact with Satan was explicit, Sperling explained in *Synopsis Physica*, but all subsequent action predicated on this pact, even unknowingly, was implicit.[65] This had been the justification for the demonization of a variety of healing and magical practices during the fourteenth and fifteenth centuries that had for centuries escaped rigorous censure. Tampering with the natural order of things by one's own free will, notwithstanding any avowed disassociation or self-protection from demonic influence, could be reinterpreted to involve the deceptive hand of Satan. However removed from its diabolical origins, the dowsing rod implied that the user remained "in league with the Devil."[66] The dowser, like many accused witches, did not necessarily meet the Evil One face-to-face and enter into a formal contract. But when dowsers and witches willfully engaged in activities that suspended the natural order of things, whether conceived as manifest or occult, and which occurred only by Satanic influence in the world, that amounted to an implicit pact.

Among the more famous promoters of Sperling's theory was the theologian and Cartesian, Nicolas Malebranche. But dowsers did not number among those men accused in Saxony or elsewhere, as Sperling and Malebranche held a minority opinion among contemporary scholars, who tended rather to remain agnostic or identify dowsing with popular superstition. The practice was not immediately harmful to the spiritual well-being of another (beyond fraud and deception), a key component of the witchcraft doctrine, and dowsers found ways to distinguish their practice entirely from crystal ball gazing, geomancy, and

[63] "Miris imposturis et mendaciis fallit atque seducit humanum genus." Sperling, *Disquisitio Philosophica*, C3.

[64] "An homunculi subterranei sint montra?" In Gerhard Heilfurth, *Bergbau und Bergmann*, 131.

[65] "Inventor primus pactum habuit Explicitum, qui sequentur hunc, media a Satana praescripta adhibendo, Implicitum." Sperling, *Synopsis Physica* (Wittenberg, 1658), 85.

[66] "Adscribimus autem totum negotium primo, qui hanc actionem tentavit...procul cum Satana iniit foedus." Sperling, *Disquisitio Philosophica*, C4.

other forms of sorcery, notwithstanding the magical books reviewed above. As a result, some dowsers paid fines, suffered the pillory, and even permanent expulsion on grounds of fraud,[67] but they never mounted the scaffold for capital punishment. Wilde concludes cautiously that treasure hunting more broadly paralleled but was never identical with witchcraft.[68]

Dowsing and Mining

High-level academic and theological support for dowsing within the context of mining preceded Sperling. The two most influential authors in sixteenth-century Western mining both worked in the famous mining town of Joachimstal—birthplace of the silver taler—on the Bohemian side of the Erzgebirge. They were the physician and humanist, Georg Agricola, and the Lutheran pastor, Johann Mathesius. Dowsing was not a major preoccupation for either, but their passing comments helped distinguish the practice from sorcery and witchcraft for many practitioners to come. Paracelsus too had great influence in mining circles, though he had little to say about treasure hunting and mining per se (aside from miners' sicknesses).[69]

Georg Agricola was highly skeptical of dowsing, but he presented a thorough and influential account of legitimate practice. In Chapter Two of *De re metallica* (1556), Agricola explained that the

---

[67] Wilde, *Die Zauberei- und Hexenprozesse*, 230.

[68] Wilde, *Die Zauberei- und Hexenprozesse*, 348, ft.57.

[69] In *De natura rerum* (1532), which, as Newman notes, may be the reworking of an authentic Paracelsian text (William Newman, *Promethean Ambitions: Alchemy and the Quest to Perfect Nature*. Chicago: University of Chicago Press, 2004. Pg 199), Pseudo-Paracelsus was skeptical of the methods commonly employed in treasure hunting. Legitimate signs of hidden treasure included images carved on the trees, marking stones, and even shadows. According to the author, the ancients buried treasure at particular times and places by the layout of shadows, and could thereby rediscover their fortune later (Sudhoff, *Sämtliche Werke*, vol. 11, 393). By contrast, the dowsing rod, which he confirmed was much in use among miners, did not offer reliable information: "Because even though they well give true indications one time, they deceive nine times, such that for every ten times, hardly one time is true [*war sagt*]" (Sudhoff, *Sämtliche Werke*, vol. 11, 393). This led the author directly to Satan's false signs, such as bizarre specters and visions at night, "against all nature." These were "pure lies and deception," he said, and appeared especially in crystal balls, mirrors, and waters, also in widespread use. In *Astronomia magna* (1537), by contrast, Paracelsus treated dowsing rods and other magical techniques as *nectomantia* (*necto*=to bind, *mando*=to order), which Paracelsus described as *Kunst*, or skilled art. In *De natura rerum*, dowsing had been merely one of the "uncertain arts…that…deceived miners."

ancients—Homer's Minerva, and Circe and Mercury—knew magical dowsing that had passed down to miners over the centuries. More recently, theologians had purified the practice of its pagan elements, though traces of ancient magic remained among "simple mining folk."[70] Like the officials we met in Chapter One, Agricola referred here casually to the 'rod' (*Rute*) rather than more derogatory 'divining rod' (*Wünschelrute*), or 'magical wand' (*Zauberstab*), whose efficacy depended on incantations and other rituals that Agricola rejected. He would not repeat the magical formula that attended the usage of magic wands, rings, mirrors, and other instruments for divining hidden knowledge of veins.[71] Focusing rather on acceptable practice, which he depicted on what became the best-known pre-modern woodcut of dowsing [Fig. 6], Agricola confirmed that the most common rod was hazel, especially that growing above mineral ore. He recounted the belief that different minerals corresponded to different woods, for example, hazel to silver bearing and pine to lead bearing ore. The practitioner grasped the rod, palms up, and paced the field "back and forth." The rod dipped only above mineral ore, an exact response to its supposed "inherent power," which Agricola did not explain.[72] The attraction was as powerful and distinct as that between lodestone and iron, barring improper usage. Among the variables determining whether a given rod worked in a particular situation, Agricola listed the size of the branch, its form, how it was handled, the power of the vein, and any inhibitory qualities of the user.

Having given the practice its due space, and having distinguished suspicious from proper usage, Agricola advised miners to prospect by other means. He allowed that dowsers could be "pious" and "serious,"[73] and also "clever" and "experienced,"[74] but the dangers far outweighed

[70] "So scheint die Rute erstmals durch das unsaubere Gebaren von Zauberern in den Bergbau gelangt zu sein; dannals fromme Männer sich von den Zaubersprüchen abwanten und sie verwarfen, wurde die Rute von dem einfachen Volk der Bergleute zurückbehalten, und die Spuren des alten Gebrauchs blieben beim Aussuchen der Gänge erhalten." Georg Agricola, *Zwölf Bücher vom Berg- und Hüttenwesen*, 32.

[71] "Der Zauberstab, mit dem die Zauberer genau wie mit Ringen, Spiegeln und Kristallen Gänge aufsuchen, kann zwar die Form einer Gabel haben, doch...nicht in der Gestalt der Rute steckt der Einfluß, sondern in den Zaubersprüchen der Lieder, die ich nicht wiedergeben darf noch mag." Agricola, *Zwölf Bücher*, 31.

[72] Agricola, *Zwölf Bücher*, 30.

[73] "Die Rute könne daher einem frommen und ernsten Manne beim Aufsuchen der Gänge von Nutzen sein." Agricola, *Zwölf Bücher*, 31.

[74] Agricola, *Zwölf Bücher*, 33.

Fig. 6. An idealized image of a dowser working with diggers and a mine surveyor. From Georg Agricola's *De re metallica* (1556).

the reputed advantages, which, in Agricola's opinion, occurred mostly by chance, in any case. Still, he refrained from giving a definitive rebuttal to silence the debate: "Because the matter is contentious and incites a great diversity of opinions among the miners, my thought is that it must be judged on its own merits."[75] Instead, he emphasized the moral and intellectual worth of the miner, and the promise of locating deposits by the surface effects of mineral vapors and other "natural indications" of ore: "Therefore the true miner, since we expect that he is a pious and serious man, does not use the magical rod [*Zauberstab*], and because he should be acquainted with and

[75] Agricola, *Zwölf Bücher*, 31.

understand the nature of things, realizes that the dowsing rod [*Wünschelrute*] is of no use."[76]

The Lutheran pastor, Johann Mathesius (1504–1565), credited his reading of Agricola with sparking an interest in mining. But Mathesius had only supportive remarks on dowsing. The Joachimstal city council called Father Mathesius—newly ordained by Luther—back to become a town pastor in 1545. His twin interests in the new theology and mining culminated in a series of sixteen sermons Mathesius delivered beginning in 1553, and first published in 1562. *Sarepta, oder Bergpostille* (*Sarepta, or Sermons on the Mount*), treated every aspect of mining and metallurgy, from the instruments of digging to minting. The author's major concern throughout was that miners live an upright Christian life. This required distinguishing between legitimate and illegitimate activities in mining, such as counterfeiting or engaging in sorcery.[77]

The dowsing rod was one example of legitimate natural magic. Mathesius did not explain the mechanism exactly, nor engage in natural philosophy. His intention was rather to present dowsing and many other techniques in mining as proper, Christian labors. Adam himself had prospected "with and without the rod."[78] It was not unlike Moses' famous staff, by which he transformed waters into blood. In the third sermon, Mathesius confirmed that the dowsing rod "turns in the miner's hand."[79] Elsewhere it was an important tool for the mine surveyor, much like the mining compass, the natural mechanism of which Mathesius also marveled at and discussed. These were some of the ways in which Mathesius, more so than Agricola, presented dowsing as a legitimate, Christian enterprise. We shall hear of Agricola's and Mathesius's works repeatedly throughout these chapters, so great were their legacies in Saxon mining.

The widespread practice of dowsing in mining was also apparent to the Catholic witchcraft scholar, Jean Bodin. While he identified all mining spirits with Satan, he echoed Mathesius and Agricola by distinguishing the dowsing rod from diabolical practice in the same demonology of 1580. The difference between legitimate and prohibited magic

---

[76] Here Agricola does refer to the rod with its more derogatory designation. Agricola, *Zwölf Bücher*, 33.

[77] See Dym, "Mineral Fumes and Mining Spirits."

[78] Johann Mathesius, *Berg-Postilla oder Sarepta: Darinnen von allerley Bergwerck und Metallen, was ihre Eigenschafft und Natur, und wie sie zu Nutz und gut gemacht, guter Bericht gegeben* (Freiberg, 1679), 30.

[79] Mathesius, *Sarepta*, 144.

of whatever sort lay in whether the cause/effect relation was common experience, Bodin argued, irrespective of whether it was perceptible (manifest). That criterion legitimized astrology, for example, insofar as the practitioner and client could determine the influence of the heavenly bodies over natural phenomena, such as the moon over the tides, even though the action itself was imperceptible. When the astrologer made judgments about a client's personal disposition or future events (judicial astrology), by contrast, the practitioner overstepped his bounds, according to Bodin.[80] There was similarly an obvious cause/effect relation between planetary phenomena and weather, between diseases like gout and its physical symptoms, and between metal and the hazel rod. And yet, the attraction between hazel and metal was "something quite commonly experienced by prospectors," even though the mechanism itself was not perceptible. Bodin added that the dowser would carry mineral ore beside his rod to augment its attractive virtue. "All these predictions are known from experience," he stressed, "although the causes are hidden and unknown. Nevertheless they are natural, and the search for them reveals the grandeur and wondrous beauty of the works of God."[81] Bodin explained that improper dowsing involved symbols, figures, and strange words, "which have no connection at all with the elements, nor with matter, nor with natural forms, nor with natural qualities." That was witchcraft.[82]

These influential works help explain why there was no concerted persecution of dowsers or treasure hunters during the witch-hunts in France or Germany. We might add that widespread interest in such harmless magic peaked rather *after* the craze. Bächtold-Stäubli suggests that treasure hunting increased in importance during the Thirty Years War when money and jewels were more frequently buried or otherwise hidden from plunder.[83] Eva Labouvie argues that interest in material-oriented magic such as dowsing in the Saar region paralleled the growing importance of private property in the eighteenth century.[84] William Hagen suggests that prosecution of such magic in Prussia increased as the state wished to curb the risk-taking mentality that

---

[80] Jean Bodin, *On the Demon-Mania of Witches*, Randy Scott, tr. (Toronto: Center for Renaissance and Reformation Studies, 1995), 80.

[81] Bodin, *On the Demon Mania*, 84.

[82] Bodin, *On the Demon Mania*, 86.

[83] Bächtold-Stäubli, *Handwörterbuch*, s.v. "Schatz."

[84] Labouvie, *Verbotene Künste*, 300.

treasure hunting represented.[85] In the following section, we introduce another reason dowsers escaped suspicion of sorcery and witchcraft: a theory of mineral vapors.

Dowsing and the Witterungen

Mineral vapors had a long history in natural philosophy before explaining dowsing. Avicenna (*Abu Ali ibn Sina*), among other late medieval Arabic philosophers, commonly identified the two underground vapors that Aristotle had spoken of in his *Meteorology* (one dry, one moist) with sulfur and quicksilver, respectively. The Arabic texts also claimed that the two principles were compositions of the four classical elements, and that they were instrumental, under influence of planetary bodies and other factors, in the generation of minerals and metals. The sulfuric and mercurial fumes exuded upwards and congealed into various earthy materials. In his *On Minerals* of around 1260 (first printed in 1476), the Dominican, Albertus, built on his predecessors. He explained how mineral vapors, under the influence of the stars, seethed upwards through the cracks and fissures to gradually congeal, depending on local conditions.[86]

The notion that the earth produced fumes was not, of course, specific to natural philosophy. *Witterungen* ('weathers') could refer to various atmospheric fogs, hazes, miasma, and fumes in more everyday usage, and in the specific context of mining, they were rushing underground vapors of mineral origin. In *De natura rerum*, Pseudo-Paracelsus described *Witterungen* much like sparks in the mining shafts, and distinguished three varieties: white lights signified silver, tin, or lead; red lights signified copper or iron; and yellow lights signified gold. He added that subtle flashes indicated fine and rich metal, whereas more blinding flashes suggested gross and imperfect metal. All varieties were sure, God-given signs of ore.[87] The author also spoke

[85] William Hagen, "Glaube und Skepsis eines magischen Schatzgräbers: Ein Fall aus der Prignitz und Mecklenburg aus den 1760er Jahren," *Historie und Eigen-Sinn: Festschrift für Jan Peters zum 65. Geburtstag*, Axel Lubinski, Thomas Rudert, and Martina Schattkowsky, eds. (Weimar: Verlag Hermann Böhlaus Nachfolger, 1997).

[86] Dorothy Wyckoff, "Albertus Magnus on Ore Deposits," *Isis* 49:2 (June, 1958), p. 122. Also see John A. Norris, "The Mineral Exhalation Theory of Metallogenesis in Pre-Modern Mineral Science," *Ambix* 53:1 (March, 2006), pp. 43–65.

[87] Karl Sudhoff, *Paracelsus: Sämtliche Werke. I Abteilung. Medizinische, naturwissenschaftliche und philosophische Schriften* Vol. I–XIV (1922–1933), vol. 11, 393.

of the "hot dry spirit" and "warm mineral heat" that seethed upwards, "like boiling water in a kettle."[88] It might appear like a smoke on the earth's surface, and it thwarted growth, evaporated dews, and melted snows. The author called these "natural signs" by which mineral veins were revealed.[89] Agricola claimed that mineral warmth melted frost, gave leaves a bluish or lead-like color in spring, darkened branches and stems, and dried roots, which frequently resulted in the toppling of trees. These were the 'natural indications' of underground mineral veins.[90] Mathesius also accepted that *Witterungen* were a good sign of mineral ore. Just as God made fire appear in the desert to punish the Israelites, so the mines exuded similar heats and flames.[91] Mathesius added that "skilled diggers mind the trees" for withered leaves and branches, stunted and twisted growth, dried roots, and a particular sponge or fungus, all effects of steady mineral heat.[92] He also noted that the miners' candle often ignited underground steams and miasmas.[93]

The first text to tie mineral vapors to dowsing practice may have been Johann Thölde's *A Report on Mining* of 1600. The subtitle read: "How to Mine by Means of Rods and Vapors." Klaus Priesner identified Thölde (~1565–1614), an overseer of salt works (*Pfannerherr*) and head overseer of mines in Bamberg (then a Bishopric), as one author behind the pseudonym 'Basil Valentine.'[94] He was best known for texts on alchemy and medicine, including a treatise on antimony. 'Valentine' was the supposed author of a five-part treatise on mining, minerals, alchemy, and alchemical medicine called the *Last Testament*, first published in 1626. It may have been a compilation of earlier works, although Part One, which is the *Report on Mining*, was likely Thölde himself.[95]

[88] Sudhoff, *Sämtliche Werke*, vol. 11, 391.
[89] Sudhoff, *Sämtliche Werke*, vol. 11, 391.
[90] Agricola, *Zwölf Bücher*, 33.
[91] Mathesius, *Sarepta*, 142.
[92] Mathesius, *Sarepta*, 143–4.
[93] Mathesius, *Sarepta*, 554.
[94] Claus Priesner, "Johann Thoelde und die Schriften des Basilius Valentinus," in Chrisoph Meinel, ed., *Die Alchemie in der europäischen Kultur- und Wissenschaftsgeschichte* (Wolfenbüttel: Herzog August Bibliothek, 1986). See also Gerhard Lenz, *Johann Thölde, Paracelsist und Chymikus und seine Beziehungen zu Landgraf Moritz von Hessen-Kassel*, Dissertation Manuscript, University of Marburg, 1981.
[95] Priesner identifies Part Four of *Last Testament* (the *Haligraphia*) with Thölde, and Priesner's argument for this authorship applies for Part One as well. Claus Priesner and Hans-Henning Walter, eds., *Johann Thölde: Haligraphia, Beschreibung aller Saltz-Mineralien und Saltzwercke* (Freiberg: Drei Birken Verlag, 2008), 125.

One Elias Montanus had already published this part in 1600, and this was probably the original Thölde.[96]

The text deserves some attention here, since its long-term influence in mining will become obvious. Thölde began by developing a unity of celestial and terrestrial bodies, tying minerals and mineral generation to larger astrological, alchemical, and Christian beliefs. God communicated His wisdom and plan to mankind through creations both above and belowground. God had created metals at Creation, forming them like metalline fruit from water.[97] Like other living substances, metals possessed a principle of fertility or life (*Ferch*).[98] This was a life principle that actualized and eventually died in cyclical process of generation and decay. The Ferch, working with the twin powers of mercury and sulfur, hardened or coagulated metallic juices into form. In its purest metalline form, the Ferch existed in a state of rest. But when decaying from perfect form, and thereby losing the properties of metal, the Ferch had awakened and begun consuming its own body. This cyclical process of growth and decay resembled the rotation of planets, which of course had a close association with the metals.[99]

The unity that Thölde posited between planetary motions and subterranean phenomena involved the inhalation (*Einwitterung*) and exhalation (*Auswitterung*) of metals, literally their breathing: "The *Witterung* is just like a breath that goes in and out of a person,"[100] although it was less visible than actual human breath.[101] Mineral ores exuded distinct fumes in the course of generation and decay. Classical and Arabic theorists spoke of heats and vapors, but Thölde's and other mining books gave the *Witterungen* a specific form and function. According to Thölde, mineral breath was an invisible "fiery aerial heat" (*fewrige Luffthitze*)

[96] Elias Montanus, *Bergwerckschatz, das ist, außführlicher und vollkommenener Bericht von Bergwercken, nach der Ruten und Witterung künstlich zubawen* (Frankfurt am Main, 1618).

[97] Thölde, *Letztes Testament*, 118.

[98] The Grimm Brothers identified '*Furche*' as a furrow-like formation in the earth. Perhaps Thölde is referring as much to the watery material that gathered in underground *Ferche* as to the formation itself. See Grimm and Grimm, *Deutsches Wörterbuch*, s.v., "Furche." It is similar to the *Archaeus* about which Paracelsus speaks in *Book of Minerals*, chapter nine.

[99] Montanus [Thölde], *Bericht*, 99.

[100] Montanus, *Bericht*, 89.

[101] "Wenn einen Menschlichen Athems kan man ja noch besser vernehmen denn einer Witterung muß man sie an einer Rutten erkennen." Montanus, *Bericht*, 56.

that passed between the heavens and earth.[102] By inhaling these fumes, the Ferch nourished itself, and it inhaled so long as it remained in the process of generation. The major source of nourishment (both above and below the ground) was the sun, which cast its fumes or rays through the stars and earth. The Ferch in turn drew the sun's fumes below the earth's surface.[103] The mineral ore slowly digested the nourishing juice or sap that it pulled downward until it became a part of its metallic body, and in this way the solar fumes promoted the generation of metal. Just as terrestrial vapor fell to the earth in the form of rain, snow, and frost, so did these vapors descend as a "metalline rain."[104] When the Ferch began to die, and consume its body, it released progressively stronger *ex*halations upwards, until the entire body was dead.

The dowsing rod responded to the back and forth breathing of superior and inferior fumes. This was a natural and immutable mechanism divorced from particular inclinations of the dowser: "Whoever deals with rods must not proceed according to his own fantasy, or bring novelties into mining...he must rather learn it from nature, and concerning rods, heed the *Witterung*."[105] Thölde believed that practitioners had bastardized its proper use: "The dowsing rod has fallen into abuse... and thereby become ignoble."[106] Many theorized that the characteristics of a dowser had bearing on the efficacy of his rod, but Thölde felt that commoners and miners had misinterpreted the rod's natural functioning by appealing to idiosyncrasies and popular magic:

> In the name of God they go, and if it [the rod] dips, then it is good, but if not, then their unlucky hand or misfortune is to blame...thus people are deluded into thinking that their own skill either hinders or promotes it rather than God's gifts and blessing; and the bulk of them... wear it under their belt, or on their hat, and protect it as sacred, each according to his own superstition."[107]

---

[102] Montanus, *Bericht*, 57.

[103] "...die führet die Spaise zu, und treibet alles unter sich dem Metall zu...doch nicht ferner denn auff die Erden, aber durch die Thamerden...da nimbt dann das Ertztes oder Metalls Einwitterung...dasselbige zu sich, und brauchet und nützet es." Montanus *Bericht*, 55.

[104] Montanus, *Bericht*, 97.

[105] "Der da mit Rutten umbgehet, der muß nicht seiner Fantasey nach gehen, und... aus seinem Sinn in das Bergwerck was newes bringen...sondern er muß sie von der Natur lernen, und also was die Rutten belanget, auff die Witterung achtung geben." Montanus, *Bericht*, 78.

[106] Montanus, *Bericht*, 81.

[107] "...und gehen also auff Gottes berath...dahin, schlägt sie ihn, so ist es gut, schlägt sie aber nicht, so ist es seiner Unglück hafftigen Hände oder Unglücks schuld...denn auß falschen wahn, meinet der Mensch immer seine Geschicklichkeit hindere

Thölde wished that such "ignorant and simple people" did not number among miners.[108]

The author distinguished seven rods: The practitioner designed the "fire rod" (*Feuerrute*) to detect inhalation that drew solar rays downward. The dowser literally set fire to the rod and entered the mineshafts, where the flame would extinguish when drawn by mineral inhalation. The "burning rod" (*Brandrute*) was covered with animal or vegetable substance, which indicated mineral exhalation by glowing.[109] A "leaping rod" (*Springrute*) was more appropriate for discovering mineralization in a perfect state of rest. It was made of two sticks that a dowser held parallel. Above mineral fumes, they separated strongly. "If it were a single stick, it would break in two."[110] Like the burning rod, the leaping rod was smeared with substance, which Thölde called 'marcasite,' defined broadly as that which increased the attraction of vapors. Each metal had its distinct marcasite—the lodestone was the marcasite of iron,[111] for example. The fourth variety, the "striking rod" (*Schlagrute*), responded to vapors that derived more directly from the Ferch itself. Thölde compared this sort of *Witterung* to the smell of a person's breath after swallowing wine. The striking rod was made of hazel and smeared with sap or fruit, on which the metallic breathing sucked, as it were. The sucking power was so strong that the rod, if released by the dowser, would stand on the ground in an upright position.[112]

The last three rods all incorporated metal. The "trembling rod" (*Beberute*) was a hollowed metal pole filled with a gold and silver amalgam.[113] The expert dowser stuck the rod into the earth, where it indicated the interaction of subterranean and solar fumes by shaking. The sixth, the "lower rod" (*Unterrute*), was a wooden branch only partially hollowed and filled with gold.[114] The metal piece responded to the downward rush of fumes, as if striving to join the solar substance with which it had an affinity. Finally, in opposition to the lower rod, the

oder fördere ihn, und nicht Gottes Gaben oder Segen, und der meiste hauffen...tragen sie doch unter dem Gürtel, oder auff dem Hut, unnd bewahren sie heilig, nach dem je einer einen grossen Aberglauben hat zu seiner armen dürfftigen Hand." Valentine, *Letztes Testament*, 80.

108 Valentine, *Letztes Testament*, 85.

109 Montanus, *Bericht*, 82.

110 Montanus, *Bericht*, 85.

111 We should therefore not assume that Thölde's 'marcasite' was iron sulfide, the modern-day designation.

112 Montanus, *Bericht*, 90.

113 Electrum is a gold-silver alloy

114 Montanus, *Bericht*, 96.

"upper rod" (*Oberrute*) drew underground vapors upwards, almost like a fishing pole caught fish. It was made of hazel and contained a small amount of "mercury of metals."[115] Thölde must have meant the mercury of the alchemists. The diviner used an upper rod to estimate the depth of ore, and to determine the type of mineral: a strong dip meant lead, whereas a weaker one could mean tin or bismuth; an even weaker dip suggested copper or iron.

'Valentine' had great influence among mining officials across Europe well into the eighteenth century. The name will echo through the mining books and other materials we review in subsequent chapters, just as resoundingly as Mathesius's and Agricola's. Mineral vapor theory was the single most important interpretation of dowsing before electricity provided a new conceptual framework around the turn of the nineteenth century.

The Rod Defended

Dowsing with a forked stick was an ancient practice that blurred the divide between legitimate and prohibited magic. Forces of social disciplining and confessionalization came to the mines, and Luther denounced all treasure hunting, but dowsing did not suffer the same fate as mining spirits and other popular beliefs did. In *De re metallica*, the skeptical humanist, Agricola, simply did not know what to make of the great popularity of the practice. He saw that dowsing rods were sometimes used in sorcery, even black magic, but he recognized that simple God-fearing miners also embraced the rod, as Father Mathesius made clear in his sermons. The witch-hunts sparked a more formal defense of dowsing, in particular within the important mining industry. Even Bodin, who did much to define the witch, positioned dowsing firmly on the side of legitimate natural magic, and Thölde extended the theory of mineral vapors to account for the motions of the rod. In the following chapter, we see that mining officials after 1650 presented dowsing in their mining books as a study of mineral vapors.

---

[115] Montanus, *Bericht*, 100.

CHAPTER THREE

MINING SCIENCE: VERNACULAR KNOWLEDGE

> Report on what special knowledge [*Wissenschaft*] is obtained from the divining rod in the furthering of noble mining:
> In the business of mining, it is undeniable that God the Highest in His grace has granted to some who make this their profession, more than to others, special knowledge, such that much is learned through nature and experience to promote mining. Although, indeed, mineral rifts and veins can reveal themselves by their outcroppings, hidden veins, by God's grace, are revealed with the divining rod, and anyone for whom the divining rod dips by nature can learn much in this manner.[1]
> –*Anonymous eighteenth-century manuscript*

To historian of mining, Hans Baumgärtel, divining and other miner traditions constituted a 'side branch' in the history of mining science. The major theoretical influence in mining, in his view, was the mechanical philosophy of Descartes and Boyle, which unified with production at schools like the Freiberg Academy in the eighteenth century.[2] However, dowsing was a central concern in the anonymous report above, preserved in the Saxon State Archive with other mining-related documents of the early eighteenth century. The report addressed major questions concerning the *Wissenschaft* of dowsing, including whether it required holding a piece of lodestone, whether the rod dipped to all varieties of mineral earth, including saltpeter, cobalt, and cinnabar, and whether it distinguished between minerals. In this report, '*Wissenschaft*' meant the God-given ability to acquire hidden knowledge of nature, clearly a different understanding of the term than Baumgärtel's.

This chapter focuses on the mining knowledge (*Berg-* or *Montanwissenschaft*) dominant in Saxon mining during the economic rebound

[1] "Nachricht, was durch die Wünschel Ruthe vor sonderbahr Wißenschafft bey Bauung des edlen Bergwercks zuerlangen." DHSA, Loc. 10690/47, ff. 33–34. Undated.

[2] Hans Baumgärtel, *Vom Bergbüchlein zur Bergakademie. Zur Entstehung der Bergbauwissenschaften zwischen 1500 und 1765/1770* (Leipzig: Deutscher Verlag für Grundstoffindustrie, 1965), 77.

that followed the Thirty Years War. It argues that, notwithstanding a growing bureaucratic and scholarly presence in mining, and substantial top-down reforms, mining officials like Nicolaus Voigtel (Chapter One) advanced a vernacular, or experience-based science. Their mining books (*Bergbücher*) upheld a new standard of systematization and professionalism that was not incompatible with miner lore, an (al) chemical conception of mineral generation, and digging and dowsing knowledge. To these officials, *Bergwissenschaft* was a synthetic body of knowledge of disparate social and intellectual origins, including the mathematics of mine surveying, and craft skill and tacit knowledge such as dowsing. It is simply not true that mathematics and theory informed mining practice only after Descartes.

To appreciate the mining science that preceded the Mining Academy, we must discard the notion developed by some economic historians and historians of geology that the crafts were devoid of theory prior to the eighteenth century. For example, though Smith and Forbes conceded that Boyle, Glauber, and other natural philosophers had an interest in artisanal knowledge, the authors argued that artisans and scholars remained distinct social groups: "The great sixteenth-century works on metals have many quantitative aspects derived from experience, but make no effort to elucidate theory. To discern any theoretical metallurgical considerations of lasting importance we must await the eighteenth century."[3] And also: "Practical knowledge continued far in advance of theory. Not until the eighteenth century was theoretical science in a position to aid practice in any but minor ways."[4] Mining historian Hermann Kellenbenz adopted the same rigid distinction between theory and practice for the pre-industrial period: "Technologically speaking, the age to which the great scholars Copernicus, Galileo, and Newton belonged was not stamped by science and learning but by the work of practical men."[5] Thomas Kuhn discussed an older tradition of mathematical inquiry and newer tradition of experimental work, which, he argued, remained divorced before the nineteenth century.[6]

[3] Cyril Stanley Smith and Robert James Forbes, "Metallurgy and Assaying," in *A History of Technology*, vol. 3, Charles Singer, ed. (New York: Oxford University Press, 1954–84), 56.

[4] Smith and Forbes, "Metallurgy and Assaying," 68.

[5] Kellenbenz, "Technology in the Age of the Scientific Revolution," 264.

[6] Thomas Kuhn, "Mathematical versus Experimental Traditions in the Development of the Physical Sciences," in *The Essential Tension* (Chicago: University of Chicago Press, 1977).

Roy Porter contended in *The Making of Geology* that, "practical involvement with the Earth—farming, prospecting, mining, surveying—formed no springboard for scientific inquiry."[7]

Yet, as Braunstein observed over two decades ago, "there was an intermediary culture between science and practice, a sort of general know-how which was present in every metal workshop and plant."[8] Pamela Smith argues similarly that craftsmen advanced a form of theorizing in their workshops, if not the academic sort that historians have emphasized. Paracelsus in particular provided an 'artisanal epistemology' of great significance to the crafts.[9] We see below that mining officials drew on Paracelsus, Johann Thölde (Valentine), Mathesius, and available mining books in explaining the origin and location of mineral ore in particular, and how to find it.

The introduction addressed the mining boom of 1450–1550. There were three major periods of growth in early-modern Saxon mining: the early boom, then a period of slow recovery after the Thirty Years War (1618–1648) that lasted roughly to the Seven Years War (1756–1763), and lastly, another wave of reform, state management, and technical advancement that began after 1763, when bureaucrats and officials founded the Freiberg Mining Academy. The following concerns the second period of steady growth.

Mining Rebound and Earth Science at Freiberg

The Thirty Years War had devastating effects on Saxon mining. Already competing unfavorably with American silver, Saxony witnessed armed conflict between 1631 and 1635, and again between 1639 and 1645. Investment and production almost stopped at many mining centers, as imperial and Swedish troops plundered villages, epidemics of plague spread, and inhabitants fled. The population declined by almost half. In Schneeberg, only one house out of every six was left standing by war's end, and at Freiberg, a population of over 12,000 in 1600 fell to some 6,500. An Imperial army occupied and pillaged the town in 1632, and the Swedes besieged it both in 1636 and 1642. Between 1545 and

[7] Roy Porter, *The Making of Geology: Earth Science in Britain, 1660–1815* (New York: Cambridge University Press, 1977), 23.

[8] Philippe Braunstein, "Innovations in Mining and Metal Production in Europe in the Late Middle Ages," in *Journal of European Economic History* 12:3 (1983), 577.

[9] Smith, *The Body of the Artisan*.

1560, average yearly production of silver in Saxony was 26,300 kg; between 1601 and 1620, it had fallen to 10,200 kg; and between 1641 and 1660, it was down to 5,200 kg.[10] Currency was devalued in an attempt to counter rising prices, and a series of minting ordinances attempted to exert centralized control from Dresden. A flourishing black market and the rule of counterfeiters characterized what economic historians term a period of 'trimming' and 'weighing' (*Kippen und Wippen*) of coins.[11] This was the great 'Decline' (*Abnahme*) that Johann Deucer, preacher at Schlackenwald on the Bohemian side, feared in a 1612 sermon. Among the major causes for the decline of mining, he listed war, rising prices, and the death of experienced miners, all of which hit the region before 1648.[12]

After the war, the mines recovered for a number of social, political, and economic reasons. Average yearly production of silver in Saxony between 1661 and 1680 was 5,900 kg; between 1701 and 1720 it rose to 9,750 kg; and between 1721 and 1740 it reached 12,950.[13] Freiberg achieved pre-war levels by 1680.[14] It smelted 2,640 kg of silver on average each year between 1649 and 1676, reached 3,510 kg between 1683 and 1695, then 4,540 kg between 1696 and 1720, before smelting a remarkable 5,610 kg in 1716 alone.[15] Two technical causes for the upswing was the use of gunpowder after 1643 to blast rock, and mine surveyor Balthasar Rößler's new hanging compass of 1633. An important social cause was the fleeing of Protestants from re-Catholicization in Bohemia. They settled in the Ore Mountains throughout the war, boosting the labor force. In 1654, Elector Johann Georg I founded the mining town of Johanngeorgenstadt for one hundred of these immigrant families. In addition to cobalt and bismuth, some 30,000 kg of silver were extracted here between 1658 and 1720.

Dresden took increasing control over juridical and police matters in mining, provided more financial support, and appointed new officials. The reenactment of mining privileges and encouragement of 'free'

[10] The Kölner Mark was standard in Germany between 1570 and 1857, according to which One Mark=234g. See Baumgärtel, *Bergbau und Absolutismus*, 103.

[11] *Kippen=abschneiden* and *Wippen=wägen*. See Uwe Schirmer, "Die wirtschaftlichen Wechsellagen im mitteldeutschen Raum (1480–1806)," in *Leipzig, Mitteldeutschland und Europa: Festgabe für Manfred Straube und Manfred Unger zum 70. Geburtstag* (Leipzig: Sax-Verlag Beucha, 2000).

[12] Johann Deucer, *Zwo Christliche Bergpredigten*, 30.

[13] Baumgärtel, *Bergbau und Absolutismus*, 13.

[14] Schirmer, "Die wirtschaftlichen Wechsellagen," 314–5.

[15] Schirmer, "Die wirtschaftlichen Wechsellagen," 319–320.

digging (*Bergfreiheit*) stimulated production. In one of his many mining ordinances, mandates, patents, and decrees, Georg I clarified miner freedoms and called for better conditions and higher yields in 1624.[16] After the war, he issued legislation designed to promote the renovation of destroyed mines and smelting works. In 1653, he stipulated that miners enjoy their traditional freedoms and protection from landowners in the field, and in 1654 he sought to remedy the lack of housing and the minimal production at the mines, since everything was "burnt out, torn up, and destroyed."[17] Other mining and minting ordinances followed, as the Electors attempted to pull unofficial coins out of circulation (especially at the Leipzig fair) and standardize the currency, goals that also demanded an efficient bureaucracy. In 1659, Johann Georg II (r.1656–1680) issued a decree based on the report of a committee he charged with investigating and recommending improvements to mining. Among other points of administration and finance, it stressed the importance of strengthening the hierarchy in Dresden and Freiberg: "We have…noted that, as yet, not all due respect has been afforded to the central and local administration by insubordinate officials and miners." Article Two proceeded to require formal apologies and recommend heavy penalties, and even imprisonment for such insubordination.[18] In 1661, the Elector organized an independent mining commission (*Bergratskollegium*) in Dresden to oversee and direct all mining, smelting, hammering, minting, tax collection, and wood and river use, and to mediate between regional officials and the Elector.

The head overseer of the Mining Office (*Oberberghauptmann*) directed the thirteen Saxon mining jurisdictions from Freiberg.[19]

---

[16] 'Bergwercks-Decret, oder Abschied, die Abschaffung oder Erleichterung der Bergwercks-Gravaminum, auch ertheilte neue Berg-Freyheiten betreffend, den 17. Mai Anno 1624,' in W.M. Schaffrath, ed., *Codex Saxonicus: chronologische Sammlung der gesammten praktisch-gültigen Königlich Sächsischen Gesetze von den ältesten Zeiten, vom Jahre 1255 an bis zum Schlusse des Jahre 1840* (Leipzig, 1842), 264.

[17] 'Rescript, die Gerichtisbarkeit über die Hut- und Zechenhäuer betr. vom 30 April 1653' and 'Befehl, den Wiederaufbau der Hut- und Zechenhäuser und die Gerichtsbarkeit über sie betr., vom 1. September 1654.' *Codex Saxonicus*, 307, 309.

[18] 'Bergwercks-Decret, oder Abschied, wodurch denen in dem Erz-Gebürge bey dem Bergwercks-Bau eingerissenen Mißbräuchen abgeholffen worden, den 6. August Anno 1659.' *Codex Saxonicus*, 320.

[19] The Saxon Central Mining Office (*Oberbergamt*) at Freiberg dates to 1554 when the Elector of Saxony created a head overseer of mine managers (*Oberbergmeister*). In the eighteenth century, the administration would expand to encompass juridical, police, and fiscal functions in mining. The Mining Academy became an arm of the Office.

Abraham von Schönberg filled this high office in 1676 and spearheaded a number of programs and reforms. He swiftly eliminated an age-old source of contention by mapping all jurisdictional borders. Under his direction, the mining administration was housed in a stately new office building in Freiberg. Schönberg's 1693 *Ausführliche Berginformationen*, a manual for officials and investors, clarified the functions, duties, and rights of all positions from manager to hacker underground, and the rules of private investment. In addition to his work in mining, Schönberg also promoted the new porcelain industry at Meissen, and oversaw the construction of a 24-by-225 meter, two-level bridge in 1685 (the *Altväterbrücke*), a demonstration of the sophistication of Freiberg craftsmen under his leadership: one level served to transport water, and the other pedestrian traffic.[20]

Friedrich August I (the 'Strong') assumed the electorate in 1694 and became King of Poland three years later. He initiated even more rapid bureaucratic expansion and implemented mercantilist policy in his forty-year reign, including overseeing a new silk industry and founding a College of Commerce in 1703 to supervise all production and trade. When Tsar Peter I stopped in Freiberg in 1711 en route back to Russia, silver production was increasing by some 4.5 tons per year. He visited the 'King August' shaft, which recalled both Elector August of the sixteenth-century silver rush and the current Elector.[21] At the Dresden court, mining was represented in processions and festivals as the "official state industry."[22]

A revived interest in the structures and properties of the earth among scholars and state officials like Schönberg accompanied these bureaucratic and court developments, as acquiring specific knowledge of subterranean things was a state interest. Naturalists and physicians across Europe became greatly interested in the 'natural riches' of their respective territories, as cameralist economic theories prevailed. These authors cultivated a new genre of regional mineralogy, as caves and

---

[20] Wolfgang Jobst and Walter Schellhas, *Abraham von Schönberg—Leben und Werk: Die Wiederbelebung des erzgebirgisches Bergbaus nach dem Dreißigjährigen Krieg durch Oberberghauptmann Abraham von Schönberg* (Leipzig: Deutsche Verlag für Grundstoffindustrie, 1994).

[21] Kasper and Wächtler, eds., *Geschichte der Bergstadt Freiberg*, 154.

[22] Helen Watanabe-O'Kelly, *Court Culture in Dresden from Renaissance to Baroque* (New York: Palgrave, 2002), 123. On August the Strong, Karl Czok, *August der Starke: Sein Verhältnis zum Absolutismus und zum Sächsischen Adel* (Berlin: Akademie Verlag, 1991), and his *August der Starke und Seine Zeit: Kurfürst von Sachsen, König in Polen* (Leipzig: Edition Leipzig, 1989).

mines lost their traditionally taboo status among the learned, and collecting minerals, rocks, and other 'fossils' became a trademark of the savant.[23] Freiberg, already home of the Central Mining Council, developed into an international center for mining-related inquiries, where scholars and officials would play a major role in the development of mineralogy and geology.[24] A number of formal and informal developments occurred that forged new institutional patterns and social bonds among natural philosophers and mining officials that would culminate in the Freiberg Academy itself. Schönberg was crucial on social and intellectual fronts. He was acquainted with the polymath, Ehrenfried Walther von Tschirnhaus, who formed a society of natural philosophers in the 1690s, and Schönberg likely met with Leibniz, who visited Freiberg in 1680 and again in 1688. In 1702, Schönberg convinced Elector August I to create a stipend fund (*Stipendienkasse*) of 300 gulden yearly for local students, intended "toward the learning of mining sciences, smelting, and mine surveying." The Elector also wished Freiberg students to gain experience at foreign mines when he renewed the stipend in 1709.[25] Mining historian Hans Baumgärtel considers this to be a decisive event in the process by which science united with production, or the landed nobility and educated elite began more systematically to apply earth sciences in mining.[26] From 1702 to 1765, there were 128 stipend-holders in Freiberg who had received subsidized training in the study of geological deposits (*Lagerstättekunde*), chemical mineralogy, measuring and mapping of veins (*Marckscheidekunde*), and other inquiries. Some of these students later became original instructors at the Academy, proposed by Schönberg in 1712 and founded in 1765.

One famous personality offering informal instruction from his home in Freiberg was the town physician and student of Georg Ernst Stahl's, Johann Friedrich Henckel (1679–1744). Among his best-known works was a 1722 text on the similarities between the plant and

[23] Alix Cooper, *Inventing the Indigenous: Local Knowledge and Natural History in Early Modern Europe* (New York: Cambridge University Press, 2007), especially Chapter Three. For the later eighteenth and nineteenth century, see Nicholas A. Rupke, "Caves, fossils and the history of the earth," in Cunningham and Jardine, eds., *Romanticism and the Sciences* (New York: Cambridge University Press, 1990), and Hamm, "Unpacking Goethe's collections."

[24] Lauden, *From Mineralogy to Geology*, 48.

[25] "Zu Erlernung der Berg-Wißenschaften, Schmeltz- und Marckscheider-Kunst." Quoted in Jobst and Schellhas, *Abraham von Schönberg*, 49.

[26] Baumgärtel, *Vom Bergbüchlein zur Bergakademie*, 172.

mineral realms, a 1725 work on pyrite (*Kies*), and a more general textbook on mineralogy of 1747 entitled, *Unterricht von der Mineralogie* (*A Lesson on Mineralogy*).[27] In his *Lesson,* Henckel began with water, describing a branching structure, much like the human circulatory system, that extended to the earth's center: "These [waters] come out from the depths of the earth, and are like veins or canals from the great world ocean, like from a heart in circulation."[28] Henckel believed that all minerals developed organically from an original watery substance at Creation. The theory had its origins in the alchemical tradition, and more recent authors like Jean Baptiste van Helmont and Johann Joachim Becher had developed it. A large collection of letters published posthumously in 1794 shows that Henckel remained critical of alchemy to the end, but otherwise steeped in questions of transmutation in the mineral and plant kingdoms.[29] His final work was a translation from the French of a large alchemical text filled with astrological and cabbalistic theory, from which he claimed to derive much practical insight for chemistry.[30]

In 1735, shortly after joining the Mining Council, Henckel received state funds to expand his home laboratory for teaching and research purposes. When the chemists Michail Lomonossov and Dimitri Winogradow of the Russian Academy of Sciences at Petersburg traveled to Freiberg in 1739, they studied with Henckel. Among his other students were Friedrich Wilhelm von Oppel and Friedrich Anton von Heynitz, the future joint-founders of the Mining Academy; Andreas Sigismund Marggraf, who later joined the Berlin Academy of Sciences and headed the chemical laboratory, also the discoverer of beet sugar; and Johann Andreas Cramer, who served the Austrian and Saxon administrations. One contemporary observer described Henckel's house as, "a veritable mining academy, where various Russians, Swedes, Norwegians, Hungarians, and Germans all together…

[27] On Henckel, see Walther Herrmann, *Bergrat Henckel: Ein Wegbereiter der Bergakedemie* (Berlin: Akademie Verlag, 1962). Also *Allgemeine Deutsche Bibliographe,* s.v. "Henckel, Johann Friedrich."

[28] Johann Friedrich Henckel, *Unterricht von der Mineralogie oder Wißenschaft von Waßern, Erdsäfften, Saltzen, Erden, Steinen und Ertzten, nebst angefügten Unterricht von der Chymia Metallurgica* (Dresden, 1747), 5.

[29] Hermann, *Henckel,* 51. Henckel's posthumous letters appeared as, *Mineralogische, Chemische und Alchymische Briefe von reisenden und andern Gelehrten an den ehemaligen Chursächsischen Bergrath J.F. Henkel* (Dresden, 1794).

[30] Hermann, *Henckel,* 85.

took up his teaching.... His house was at once a hostel for foreigners and inn for the learned in mining."[31]

Henckel eventually moved his laboratory to the home of another chemist, Christlieb Ehregott Gellert. He had spent nine years at the Russian Academy and been greatly influenced by Lomonossov before returning to Freiberg in 1743.[32] Gellert had earned an international reputation for his knowledge of metallurgical chemistry, and after Henckel's death, his younger colleague continued instruction, charging an elite clientele up to 400 taler for his service. Gellert was also the first to implement the amalgamation process at Freiberg, by which silver was extracted from its ore with mercury, rather than through high temperature fire alone. He was appointed head overseer of smelting (*Oberhüttenverwalter*) in 1762. After publishing textbooks in assaying and metallurgical chemistry, and offering years of private instruction, Gellert became one of the original Academy professors in 1765. He died a member of the Mining Council.

The *Stipendienkasse* and work of Henckel and Gellert represent a movement toward institutionalized and international-minded mining science at Freiberg. It is noteworthy that Henckel examined dowsers, but made only passing commentary on the practice in his published work. The same was true of Schönberg, who said little about dowsing in his 1693 text, but included a glossary of miner language (*Bergsprache*) at the end that included entries for '*Wünschelrute*' and '*Witterung*' (mineral vapor), and who defended hired dowsers in an official letter.[33] These men shied away from publicly endorsing dowsing before the learned public, but otherwise interacted with practitioners in their practice. We set their work into context when we study the mining books of contemporary, lesser-known authors. Here we find that *Bergwissenschaft* was a more synthetic body of theory and practice than Schönberg and Henckel described, and included tales of discovery, biblical stories, alchemical and astrological theories, dowsing theory, as well as knowledge of the mining compass (*Grubenkompass*), the mathematics of mine surveying (*Marckscheidekunde*), and the craft of mine construction. Early Academy scholars would claim to possess a *Wissenschaft* that rendered much of this knowledge obsolete.

---

[31] In Werner Lauterbach, *Bergrat Christlieb Ehregott Gellert* (Leipzig: Akademie Verlag, 1994), 47.

[32] Lauterbach, *Bergrat Gellert*, 38.

[33] On Henckel, Schönberg, and the patronage of dowsing see Chapter Four.

But here we see how robust the vernacular science was that preceded the Academy.

Digging through Mining Books

Mining officials had produced manuals and texts pertaining to various facets of the enterprise for centuries. The generic title 'mining book' (*Bergbuch*) does not distinguish between the many types of texts and their audiences, both manuscript and printed, including tables of extraction figures, shareholder lists, mining 'customs' and laws, minting manuals, assaying manuals, manuals on the measuring and charting of veins, glossaries of mining and metalworking terms, chronicles, general overviews of the administration and procedures of mining and metallurgy, and all manner of compilations. Johann Mathesius printed a collection of 'mining sermons' that covered standard practical information on mining and metallurgy. It too functioned as a practical mining book in Saxon mining. Some of the earliest printed manuals often cited in the literature are the Freiberg physician and mayor Ulrich Rülein von Kalbe's *Bergbüchlein* of 1500, an instructional dialogue between an expert and novice miner intended to attract investment in Saxon mining; the Siena minter Vannoccio Biringuccio's metallurgical treatise, *Pirotechnia* (1540); Ludwig Lässl's ornate manuscript, the *Schwazer Bergbuch* (1556), a general overview of mining in the Tyrol; and Lazarus Ercker's metallurgical manual, *Aula subterranea* (1574). Agricola's *De re metallica* (1556) was a more famous mining and metallurgical treatise, although far more scholarly and comprehensive than most other works in this genre.[34]

Economic historians and historians of mining have tended to overlook or downplay the theoretical discussions contained in these early

[34] Early mining books of more circumscribed fame include Johann Knobloch, *Der Ursprung gemeynner Berckrecht* (1535), Petrus Albinus, *Meissnische Bergkchronica* (1590), Elias Montanus (Johann Thölde), *Ein Büchlein von dem Bergwergk* (1600), Hans Uttman, *Bericht von denen Ertzgebürgen* (1601), Georg Engelhardt von Löhneyss, *Bericht vom Bergkwerck* (1617), Nicolaus Voigtel, *Geometria Subterranea* (1686), Abraham Schönberg, *Ausführliche Berginformationen* (1693), and Christoph Herttwig, *Neues und Vollkommens Bergbuch* (1710). For an overview of early German mining and metallurgical literature, see David E. Connolly, "Ulrich Rülein von Kalbe's Bergbüchlein in the Context of Sixteenth-century German Mining/metallurgical Literature," in Robert Bork, ed., *De Re Metallica: The Uses of Metal in the Middle Ages* (Burlington, VT: Ashgate, 2005), Chapter 21.

texts and manuals. While Agricola and Ercker excluded theoretical material, Biringuccio included ample discussion of alchemical theory,[35] and Kalbe was clear if succinct on his theory of metallic generation. Printers may have excised theoretical discussions in the interest of appealing to a wider audience. William Eamon studied the printed and manuscript versions of one early text entitled, 'The True Practice of Alchemy,' which underwent considerable change. The 1531 print had eliminated all "esotericism and metaphysical content."[36] The mining books we review below were largely technical in content, but substantial sections on prospecting and dowsing introduced the origin and generation of mineral veins, the structure and properties of minerals, and even the moral character of the miner. Together they offer a snapshot of mining knowledge in Saxony as the region recovered from the Thirty Years War; knowledge we then set in motion, as it were, in chapter Four.

From a great variety of mining books, we can study the *Bergwissenschaft* of three that no recent history of mining or earth science has treated: the works of one mine master (*Bergmeister*) and two mine surveyors (*Marckscheider*), all relatively high-standing mining officials. We know least about the mining master Abraham Löwel and his 1671 manuscript, *Memorial zu Erlangung einer exquisiten Bergwissenschafft* (Report Toward the Attainment of Proper Mining Knowledge). It is the carefully penned and well-bound manuscript of the highest official of one Saxon *Bergamt*, clear about his intention to build on his predecessors in grounding a new field, and steeped in mining lore, Paracelsian theory, and Mathesian religious sentiments. Historians of science focus on Agricola's humanism and identify his major influence in mining to be the technical *De re metallica*;[37] Reformation historians have treated Johann Mathesius chiefly for his role in spreading the new theology. This is in marked contrast to how Löwel distributed credit to his predecessors: Agricola provided the definitive glossary of mining words and phrases (what Löwel calls

[35] For a discussion, see Newman, *Promethean Ambitions*, 127–132.

[36] Eamon, *Science and the Secrets of Nature*, 114.

[37] Pamela Long, "Of Mining, Smelting, and Printing: Agricola's 'De re metallica,'" *Technology and Culture* 44:1 (2003), 97–101. Marco Beretta, "Humanism and Chemistry: The Spread of Georgius Agricola's Metallurgical Writings," *Nuncius: Annali di Storia della Scienza* 12 (1997). Owen Hannaway, "Georgius Agricola as Humanist," *Journal of the History of Ideas* 53:4 (1992), 553–560.

"*Informatio terminorum*")—he latinized some 600 German mining terms in 1546—whereas Mathesius produced the authoritative theory of metals and description of mining ("*Cognito rerum*") with *Sarepta*. Löwel mentioned neither Agricola's theoretical *De ortu et causis subterraneorum* nor his *De re metallica*, suggesting that these works may have found less immediate application in small-scale mining. Most of all, Löwel echoed *Sarepta* by praising the Creator of minerals, providing mining analogies for the Christian life, and including sayings from the old Joachimstal pastor himself. Complementing these two earlier works, the *Memorial* was the third branch of Löwel's proposed tripartite mining science: instructions on setting up a mine ("*Ajudicatio casuum*").[38]

We know considerably more about our second example, the man credited with improving the hanging compass (*Hängekompass*) in mining, by which the mine surveyor took angles and measurements in the field and in the tunnels below.[39] Born near Elbogen in Bohemia, Balthasar Rößler sought refuge in Saxony during the Counter-Reformation, where he was a surveyor of mines at Marienberg after 1632, a shift manager and director of an iron works, and employed at the liquation works (*Saigerhütte*) at Grünthal, before assuming a post in 1649 as mine surveyor and record keeper (*Gegenschreiber*) at Freiberg. Rößler taught mine surveying here at the fee of 60 *Reichstaler*. He died as mine manager at Altenberg in 1673. His *Hell-Pollierter Bergbauspiegel* (Brightly Polished Mirror of Mining) was printed posthumously in 1700. Rößler's book is a more formal presentation than Löwel's, divided and numbered neatly into books, chapters, and sections of paragraph length. According to Baumgärtel, the Rößler "heralded a new level of systematization in mining science."[40] Rößler dedicated his work to the head administrator in mining, Abraham Schönberg.

Finally, Augustus Beyer (1677–1763) was a mine surveyor at Freiberg in 1697, when, as a series of contracts show, he began teaching his art.[41]

---

[38] Abraham Löwel, *Memorial betreffend die Bergwissenschaf* (1671), SLUB, B158, Bl. 1.

[39] The sixteenth-century Italian physician and mathematician, Geronimo Cardano, invented a method of diminishing the motion of a compass by hanging. Rößler's adaptation for mining remained the standard into the nineteenth century.

[40] Baumgärtel, *Vom Bergbüchlein*, 61. On Rößler, see also Heinz Meixner, Walter Schellhas and Peter Schmidt, eds, *Balthasar Rösler: Persönlichkeit und Wirken für den Bergbau des 17. Jahrhunderts* (Leipzig: VEB Deutscher Verlag für Grundstoffindustrie, 1980).

[41] BAF, Loc. 3477.

Among his students were the metallurgical chemist, Henckel, as well as the joint-founder of the Academy, Oppel.[42] Beyer became superintendent of mines (*Bergkommissar*) in 1724, and he joined the city council a year later. The *Gründlicher Unterricht vom Bergbau, nach Anleitung der Marckscheiderkünst* (Thorough Instruction on Mining, with the Guidance of Mine Surveying), printed in 1749 (in manuscript as early as 1718), was one in a line of systematic, practical guides on the mathematics and instruments of measuring and mapping of mines. Unlike Löwel's, a practical manual on how to establish a mine, and intended to complement two prior works, and Rößler's, a more general account of mining in the tradition of Agricola's *De re metallica* (1556) and Löhneyß's *Bericht vom Bergwerk* (1617), Beyer's fit squarely into a genre of specialized, technical literature on a specific branch, mine surveying, that included works by Erasmus Reinhold (1574), Nicolaus Voigtel (1686), Johann Friedrich Weidler (1726), Oppel (1749), Andreas Böhm (1759), Abraham Gotthelf Kästner (1775), and many others.[43] Also unlike Löwel and Rößler, Beyer offered a more self-conscious and defensive account of dowsing, suggesting that Freiberg elites were becoming more critical of the practice in his day.

All three examples had the character of textbooks that introduced or updated a field. However paradoxical from a modernist perspective, this progressive goal was compatible with a deep commitment to the knowledge of the *Bergverständiger*. Two central components of mining science were prospecting or digging (*Schürfen*), a body of knowledge and practice steeped in mining lore and the accumulated wisdom of prospectors, and mine surveying, which relied on the mining compass and trigonometry to determine the layout of claims. The dowsing rod, to which our authors referred casually as the 'rod' (*Rute*) rather than 'divining rod' (*Wünschelrute*), found a niche in both practices.

Digging and Dowsing

All three authors listed standard surface-level treasure-hunting practices. In a chapter entitled, "On the Exposure of Veins" (*Von Entblößung*

---

[42] Herrmann, *Bergrat Henckel*, 24, 50. On Oppel, see Otfried Wagenbreth, *Die Technische Universität Bergakademie Freiberg und ihre Geschichte* (Leipzig: Deutscher Verlag für Grundstoffindustrie, 1994), 22.

[43] Augustus Beyer, *Gründlicher Unterricht vom Bergau, nach Anleitung der Marckscheiderkünst* (Schneeberg, 1749). The manuscript copy dated 1718 is housed at the Gotha Forschungsbibliothek (GFB, Ch.A. 972).

der Gänge), Rößler began with the distinction made by Agricola between discovery by means of accident and natural events (*ohngefehr*), and that by means of intentional searching (*durch Ersuchung*). Natural events included storms, mudslides, fires, winds, earth upturned by the farmer's plow, and other chance occurrences that sometimes exposed mineral ore. Rößler included mineral vapors (*Witterungen*) under this heading, the fumes believed to indicate the location of ore. Many famous mines had discovery tales involving similar natural events. Agricola rehearsed a number of stories in *De veteribus et novis metallis* (1546), and Löwel opened his text with many of the same two centuries later: Kuttenberg was founded after a monk rested his head on a chunk of mineral ore; Rammelsberg in the Harz, when a horse upturned some earth; and Freiberg, when fur traders from Bohemia, en route back to Thuringia, chanced upon silver. These and other tales coursed through sermons and chronicles before poets and philosophers of the late eighteenth century compiled larger collections of folklore. Beyer similarly distinguished between such unforeseen events as environmental and geological disturbances, or God's action, and the skill of the miner (*Bergmännischer Anweisung*) in locating mineral ore.

Turning to discovery by means of skill and exploration, which raised the issue of dowsing, again, our examples were mostly congruent. The authors began with mountain waters. Löwel echoed Paracelsus, Thölde, and other alchemical thinkers who imagined that minerals and metals were the 'fruits' of a gigantic tree of water. The experienced digger accordingly examined waters for a dusty appearance, or other such indications of mineral content. Löwel likened cloudiness in water to the smoke that filled a smelting house.[44] According to him, the miner also studied the trees for indications of mineral heats and steams (*Dünste*, *Broden*). They gave a bluish or lead-like color to leaves, blackened branches, and dried and weakened the roots, rendering trees more susceptible to toppling. Hot and sulfuric mineral heats also stunted the growth of grass and melted or evaporated snow and dew sooner than in adjacent regions not affected. Rößler cautioned, however, that mineral ore oftentimes lay too deep or in too small a quantity to exude vapors visible to the earth's surface. He recommended the

[44] "Sein die Wasser hiesig [hitzig] oder es leget sich ein Staub drauf wie ein Hütten Gestübe so giebet es eine Anzeigung daß die Gänge Metallen führen." Löwel, *Memorial*, 22.

dowsing rod in such cases.[45] Löwel also mentioned the volatile *Witterungen*, which rushed out like lightning from the mines at night, "like a blue fire."[46] Beyer began with water sources, dews, and other standard indications, including the *Witterungen*, which were "like a fire." They were different by day and night, he explained, and varied in intensity.[47]

Löwel concluded his list with comments on "*Gur*," believed to be an undifferentiated mixture of sulfur and mercury that developed into distinct metals. Mathesius mentioned this substance in his account of the generation of ore. *Gur* was similar to the 'Mercury of Metals' about which the alchemists spoke, perhaps in a natural setting, rather than artificially produced in a furnace. Said Löwel, *Bergverständiger* recognized a substance "like buttermilk" flowing from pathways in the mines and in plowed fields.[48] Zedler included an entry in his 1732 dictionary: "A humidity that flows out of stone, looks almost like buttermilk, and indicates ore."[49] The debt to alchemical thought on mineral generation could be explicit: "All true philosophers know," said Löwel, that salt, minerals, gems, and stones were the fruits of a gigantic tree of water. Water reached its "ultimate perfection" (*ultimam materiam*) in the element earth in this way, just as Paracelsus had described in *De mineralium*.[50]

Dowsing was the prospecting practice of choice, and our three officials developed the theory of mineral vapors. Bergmeister Löwel offered the most in-depth treatment, using Valentine's (Thölde's) *Bericht* and Paracelsus's *De mineralia*. In a section entitled, "Dowsing Rod: That it Dips Naturally," Löwel drew sometimes verbatim on Thölde to position dowsing in mining science. The author reviewed

---

[45] "Anderer Orthen, wo das Ertz tieff lieget, und geringes Haltes [Gehalts] ist, kan auch wohl die Auswitterung nicht so kräfftig seyn, und so dieses eine gewisse Regul, wäre viel besser darauff, denn auff die Ruthe, zu trauen, man würde sich dessen, vielfältig bedienen." Balthasar Rößler, *Speculum Metallurgiae Politissimum, Oder Hell-polierter Berg-Bau-Spiegel* (Dresden, 1700), 11.

[46] "Die Gange verrathen sich auch gemeiniglich durch die Witterung wenn nehmlich bey Nacht dieselbe sich von der warmen Lufft entzündet und einen Streichen hinaus läufft als ein blau feuer." Löwel, *Memorial*, 22b.

[47] Beyer, *Unterricht*, 8.

[48] "In allen zechen und verfahrnen Feldte richten sich die Bergverständigen nach der Guhr so aus denen Straßen gieret, und siehet wie Buttermilch welche offtmahls von Erz herfindet." Löwel, *Memorial*, 22b.

[49] Zedler, s.v. "Guhr."

[50] Löwel, *Memorial*, 14b.

how vapors had an affinity with both metals and tree branches above ground, and he associated this phenomenon with the various effects of the sun on the earth.[51] Just as the sun drew the tides, he believed, so did the rays of the sun have a similar effect on minerals. According to Thölde, as solar beams descended upon the earth, they met with the exhalation of mineral ore, which drew the beams underground, where they promoted the generation of metals. He described the push and pull of solar rays and mineral vapors as 'breathing.'[52] Holding the two arms of a forked branch, the dowser might affix gold to the extended shaft, which the solar rays and mineral exhalations in turn pulled downward, as if catching the gold in a crosswind. In Löwel's words, the *Witterungen* wished to carry the metallic piece "back home," another reference to Thölde, by which he meant that the gold originated from mineral ore. To better recognize invisible mineral vapors in his dowsing practice, moreover, the miner might smear the rod with limestone or a chalky substance (*Kalch der Erden*), which would actually illuminate when in contact with vapors.[53] Löwel likened the phenomenon to capturing smoke in cloth. According to Thölde, animal or vegetable substance covered the fire rod (*Feuerrute*) and accounted for its illumination, and other rods were covered with 'marcasite.'

Löwel also drew on Thölde when the Bergmeister claimed that the breathing of mineral vapors performed distinctive functions in nature. When a given tree grew above mineral ore, an "upper air" controlled the flow of sap or juice through the branches, leaves, and fruit, and a "lower air" drew juices into the roots. These substances, driven by upper and lower airs, traveled up and down the trees in the course of their generation and decay. They were the reason that certain branches had an affinity with and reached toward mineral ore, since mineral vapors pulled at the juices inside the branch. A freshly cut hazel stick would literally stand still over mineral ore, reported Löwel, like an iron needle placed on a magnet, propped up by rushing air. This was the

[51] Löwel, *Memorial*, 76b.

[52] "Gleich wie einem natürlichen Menschen sein natürlicher Athem in seinen Leib auß und eingehet, so ist diese Witterung auch." Montanus, *Bericht*, 86.

[53] "Weile nun solch Witterungen feuer hiezig ist, so glühet sie auf ihre weisse, und also glühet sie diesa Ruthe an, darbey mann den gewiss weiss, dass ein lebendiges Metall verhanden [vorhanden] sey, soll nun die Ruthe also angegluet werden von solchen Witterung, so muss sie ihren besondern Zeugk und Zundten [Zündung] haben, der solch ihr Feuer an sich nehme." Löwel, *Memorial*, 77.

"ore position," which, Thölde had said, occurred when mineral vapor sucked at the juicy matter inside a hazel, striking rod.[54]

Löwel ended his section on dowsing by listing more common knowledge about the practice. Rods of different materials corresponded to different minerals and metals: hazel branches to silver, ash to copper, fir to lead and tin, and even iron to gold. Alternatively, the clever dowser might affix to his rod a metal ball forged of various metals called an electrum and expect the rod to dip to mineral ores of varying content. Thölde had discussed the electrum of gold and silver that filled the hollow of a trembling rod. Löwel diverged from Thölde only in that Löwel gave the most propitious time to break a rod: the first Sunday after the new moon, at dusk, in the sign of cancer.[55] He did not explain this astrological belief, listing it simply as one more piece of common knowledge. In the *Dictionary of German Superstitions*, Bächtold-Stäubli claimed that midnight, June 24, also in the sign of cancer, was the best time.[56] Thölde ridiculed the astrological significance miners had given to the dowsing rod, believing that they corrupted a purer practice, though he allowed that solar and mineral vapors interacted in metallic generation. Both Agricola and Mathesius had taken a similar view and rejected judicial astrology in mining. Löwel reported the information, evidently common among miners, and refrained from judgment.

In *Bergbauspiegel*, Rößler first mentioned dowsing in his treatment of discovery of mineral veins by means of human skill. He noted that use of the dowsing rod predated and was far more common than tracking mineral fragments: "The method...most common in all mining towns these days, proceeds with dowsing rods; the least common with *Geschiebe* (alluvial mineral fragments), because they are useful at only a few tin and iron mines."[57] Presumably, he believed that waters ripped more easily through this ore, depositing fragments widely.

Rößler then distinguished two general theories of the rod: that it operated through a certain quality of the practitioner, and that it worked through properties of the rod itself. Rößler upheld the second

---

[54] "Wannenher ihr Gang recht stehet, das Winden aber von oben bis auf diesen Stand, da die Ruthe stehen bleibet, heist den Erz Stand." Löwel, *Memorial*, 78.

[55] Löwel, *Memorial*, 78b.

[56] Bächtold-Stäubli, *Handwörterbuch*, s.v. "Wünschelrute."

[57] "Die Ersuchung aber, so vor dessen im Brauch gewesen, und ietziger Zeit in allen Berg-Städten fast am üblichste, geschicht durch die Ruthen, die wenigste durch die Geschiebe, weil sie nur bey etlichen wenigen Zwitter-un Eisenstein-Bergwercken gebräuchlich." Rößler, *Speculum*, 11.

explanation and proceeded to undermine the authority of dowsers themselves. To him, the rarity of true practitioners, their fallibility, and the diversity of their practices proved that the rod did not operate by some virtue of man: "Among many men there is hardly one disposed [to dowse], and wherever many are used, they do not all coincide. Some find little, some find much."[58] Some rods dipped strongly, and others weakly; some dowsers worked for an hour, and others for less time; some worked with mineral veins, and others with rifts, waters, or *Geschiebe*. Rößler accordingly recommended that a miner interrogate a prospective dowser before hiring to determine his authenticity and understand his practice. He added that the practitioner might test his rod with earth samples to determine to which sort it dipped: silver bearing, iron bearing, or otherwise. Understanding the parameters of its functioning, the dowser would then pace the field more certain of his instrument's operation.

The cause of the dipping motion was the action of mineral vapors: "The exhalation or *Auswitterung* in the veins causes the rod to pull or dip."[59] If earth or stony material obstructed them in their ascension, they reached the earth's surface at some remove from their origins, up to four Lachter (8 meters) away, and with less strength. That obviously left some room for interpretation in the field. Beyond these statements, Rößler did not develop the theory of mineral vapors any further. His focus remained on practice.

Rößler also discussed the properties of the rod itself. While all sorts of trees and bushes were in use, an onion-shaped branch broken from hazel either by knife or by hand was the most common. Rößler was amazed that some practitioners were then using brass and iron rods, quite unlike in earlier times, he claimed.[60] Magnetic and electrical theories of dowsing, which involved metallic rods, would become more common. Distinguishing proper use from magic and superstition, moreover, he was adamant that a dowser use no "incantation or blessing" in the breaking of a rod. He claimed to have witnessed numerous dowsers break rods at only certain, astrologically significant times, and employ prayers in their usage. He recalled one "fraud" in particular,

---

[58] Rößler, *Speculum*, 13.

[59] "Es verursachet aber die Ruthe zum Zug oder Schlagen die Exhalation oder Auswitterung in den Gängen." Rößler, *Speculum*, 13.

[60] "Aber anizo werden auch von Meßing und Drath Ruthen gemacht, und gebraucht, davon man vor Alters nichts gewust, ist sich auch darüber zuverwundern." Rößler, *Speculum*, 13.

who specified where the next vein lay merely by raising his rod into the air like an antenna.[61] Rößler preferred the classic association, developed in Thölde and Löwel, of metals with vegetable life: in particular, seven sorts of trees with the seven known metals. All else being equal, the vapors of a given mineral ore had a natural affinity with a dowsing rod of corresponding virtue, when a skilled dowser stood directly above the ore in question.

Rößler explained how the dowser used their rods to recommend digs. He carried stakes into the field, and as he followed the supposed trajectory of a given mineral vein, he hammered a stake into the earth every 10 Lachter (20 meters) and marked it with a number or letter. Rößler depicted a dowser at various stages of this work, as a mine surveyor with measuring stick, accompanied by a hooded digger, looked on from a hilltop [Fig. 7]. The suveyor's work informed the dowser by indicating the location and direction (*Streich*) of a proposed vein—north/west, south/east—and by informing him when he had stepped into somebody else's claim, since the coordinates of all property were charted. This was especially important in the sloping terrain of the mountains, where the layout of claims became confused. Having staked out a major vein, the dowser then proceeded to search for peripheral veins by straying four or five Lachter (8–10 meters) from the central find.

Satisfied that he had exhausted a given region, the dowser then determined the most promising spots for test digs. These were the places with the highest concentration of intersecting veins (and stakes) as indicated by the dowsing rod; it was also these spots over which the rod dipped most strongly. Here Rößler warned against false signals that might come from a rotting vein, the fumes of which could be mistaken for a healthy one.[62] He also questioned the belief held by the "old," and still common among miners in his day, that two veins intersecting at a right angle—a "cross"—was a sign of rich ore. Rößler claimed, by contrast, that veins running parallel or intersecting

[61] "Ich habe unterschiedliche Ruthen-Geher gesehen, die nicht alleine gewisse Zeit zum Ruthenschneiden oder Brechen derselben, sondern auch Seegen dabey, und bey Anführung gebraucht haben: wie denn ezlicher solcher Ruthen-Geher nicht alleine haben wissen wollen, was die Gänge vor Erz führen, wie mächtig es sey, und wie viel Lachter tieff es liegt, sondern es hat auch einer, den ich wohl einen Betrüger nennen kan, wenn er die Ruthen in die Höhe gehalten, und sich mit den Leib hin und wieder gewendet, aus der Lufft vernehmen wollen, wo am nächsten Gänge anzutreffen seynd." Rößler, *Speculum*, 13.

[62] Rößler, *Speculum*, 14.

Fig. 7. A mine surveyor overlooks from a nearby hill, as a digger, directed by the dowser, hammers in stakes. From Balthasar Rößler's mining manual, *Bergbauspiegel* (1700).

gradually were more promising, whereas a violent perpendicular intersection damaged the integrity of the respective veins.[63]

These were some of the considerations of the hired dowser, according to the mine surveyor, Balthasar Rößler. Seeing that he also encouraged the "true miner" to not lose hope after an unsuccessful dig, and that Rößler repeated that mineral veins sometimes dipped below an earthy cover or were otherwise difficult to locate, we begin to understand how failures did not necessarily lead to skepticism among dowsers.[64] A student of Rößler's method would not easily lose confidence. In Chapter Six, we will meet a nineteenth-century surveyor and dowser, Karl Schmidt, who used the Rößler method almost exactly, made any

[63] "Jedoch thun diejenigen nicht allemahl gut, die, gleichsam zu rechten Winckeln, einander durchschneiden; meists verruckt, oder verdruckt einer den andern." Rößler, *Speculum*, 15.

[64] "Ein rechter Bergmann aber muß sich durch einen vergeblichen Schurff nicht abschrecken lassen, will es in einem nicht glücken, muß er es weiter versuchen, bißweilen kriechen die Gänge unter, daß ein Dach darauff liegt." Rößler, *Speculum*, 20.

number of false predictions, but continued to explain and defend his practice before a skeptical physicist, Ferdinand Reich.

In his mine surveying manual, Beyer offered the most in-depth coverage of dowsing in Saxon mining. His first mention appeared in his list of prospecting by skill (rather than luck), in which the rod was the "most common" method of discovering ore. The dowser worked closely with the mine surveyor, Beyer stressed.[65] He devoted three chapters to developing the subject. Much of the material was copied from an earlier mining handbook (distinct in style and format from Beyer's own 1718 manuscript).[66] Beyer began by referencing long-standing learned debate; texts both for and against the veracity of the practice, proving just how widely read he was on the issue. Dowsing had become a highly contested issue among scholars after the case of a diviner who discovered the culprit of a crime in France.[67] Beyer mentioned the major authors in the debate, but he did not explain or endorse the corpuscular theory the French had proposed. Among other authorities on dowsing, Beyer listed Basil Valentine and Elias Montanus (both of whom we know to be Thölde), Georg Agricola, Georg Engelhard von Löhneiß, Athanasius Kircher, Andreas Libavius, Johann Sperling, and Johann Praetorius. It would require a "large book," Beyer said, to discuss all these opinions.[68] He was forthcoming with his confidence in the practice, stressing that miners had been dowsing "continually since old times."[69] He drew a distinction between those that hastily rejected or condemned the technique, and those with greater experience of this "natural work."[70] Later

[65] "Am allermeisten aber wird Bergwerck fündig gemacht und Gänge ausgerichtet mit der Ruthe durch verständige und geübte Ruthengänger; weilen aber bey solchen ein Marckscheider viel zu erforschen und zu observiren, auch zum öfftern in vielen Aufgaben bey Ausübung der Marckscheider-Kunst mit den Ruthengänger zu thun hat, so soll von diesen in folgenden Capitel etwas weitläufftiger gehandelt werden." Beyer, *Unterricht*, 8.

[66] The earlier, anonymous handbook was, *Beschreibung des Edlen Berg-Werck-Baues, wie solches bey der Churfürstl: Sächsch: alten Freyen Berg Stadt Freyberg verführet worden und noch verführet wird. Allen denenjenigen so sich bey demselben encapiret machen wollen sehr dienlich*, SLUB R284.

[67] See Chapter Five on the Aymar case of 1692, and the corpuscular theory proposed to explain his dowsing.

[68] Beyer, *Unterricht*, 9.

[69] "...indem von alten Zeiten her solche beständig annoch bey Bergwercken beybehalten, auch viele Bergwercke darnach ausgerichtet und fortgestellet werden." Beyer, *Unterricht*, 9.

[70] "Ob nun wohl viele sind, die den Gebrauch derselben gänzlich verwerffen und vor höchst sündlich halten, so sind doch auch andere hingegen, die eine genauere Einsicht und Information davon haben, und die Wünschelruthe als ein natürliches Werck betrachten." Beyer, *Unterricht*, 9.

in the book, he censured in particular two Wittenberg scholars who publicly denounced the rod "of the honest miner" to be the work of the devil, without so much as bothering to investigate first-hand accounts.[71] These scholars were certainly Johann Sperling and Jacob Klein, who together authored the 1666 essay on dowsing we discussed in Chapter Two, which argued that the practice required an implicit pact with Satan.

Beyer discussed the acquisition of the rod and the practice itself, expressing the same practical-minded skepticism that Rößler had, and which will become typical of Freiberg officials. According to Beyer, the dowser chose a three-pronged one year-old branch, and proceeded by grasping the two shorter legs. There was a correspondence between particular persons, particular hours, particular rods, and particular metals, which the experienced practitioner understood. Beyer did not explain the system of associations in any depth, presenting it more as common sense, as Löwel had, and stressing only that the mechanism at work was natural. Beyer also accepted the theory that different metals corresponded to different dowsing rods, for example, silver-bearing ore to hazel. Among other uses, this allowed practitioners to test rods in the field to estimate the value of a find before digging. Beyer added that widespread misuse had distorted the true nature of dowsing. He ridiculed the "superstition" of breaking the branch at astrologically determined times, such as St. Johns Day (June 24) around midnight, and of using special characters, or mumbling magical words that 'bind' the rod in various ways: "Christian miners do not do such things."[72]

Beyer did not assume that miners always relied on dowsers, or that they were always reliable. His remarks demonstrate that belief in dowsing was not 'all or nothing,' but that there were degrees of acceptance.

---

[71] "...denenjenigen die die Ruthe eines ehrlichen Bergmanns als ein Teuffels-Werck ohne vorhergegangene hinlängliche Untersuchung auch zu unsern Zeiten in offentlichen Schrifften und besonders in zweyen Wittenbergisches Disputationibus schlechterdings verworffen haben...denenjenigen Herren Gelehrten aber die dergleichen Sachen untersuchen, überlasse nachstehende Begebenheiten zur weitern Untersuchung ihrer Ursachen." Beyer, *Unterricht*, 221.

[72] "Daß aber solche zu gewissen Zeiten in der Johannis- oder Christnacht zwischen II. und 12. Uhr, und darze nackend und mit Sprechung gewisser Worte, als In principio erat verbum etc. oder mit gewisser Characteren bezeichnet müsse geschnitten werden, ist ein Aberglaube und gottloser Mißbrauch; Christliche Bergleute thun solches nicht, und binden sich weder an die Zeit, noch an einige Worte, sondern gebrauchen sie, wenn sie nöthig ist, und wo sie Holz, auch Gelegenheit zu schneiden haben." Beyer, *Unterricht*, 10.

For example, he claimed that the practice of holding samples of mineral ore beside the dowsing rod was new to practice and highly dubious, in his words, "very uncertain and without solid grounding."[73] Moreover, he recognized that there were as many methods as there were dowsers. For some practitioners, the rod dipped above topsoil; for others, it dipped over grass; and still others worked only in stony regions. Beyer recommended consulting more than one dowser before committing to a dig. Some dowsers were too sensitive, and their rods dipped to crevices and gaps as if to mineral veins. Beyer suggested having more than one rod check the same spot.[74] More importantly, the dowser's claims did not always agree with those of the higher-standing mine surveyor (*Marckscheider*). In that case, Beyer recommended that the surveyor be trusted first; when the two were in agreement, however, the miner or investor had more reason to believe.[75] Beyer concluded that it was "undeniable" that a properly used dowsing rod produced useful knowledge, but to rely entirely or uncritically on this method was ill advised, as it was "subject to many hindrances and accidents."[76]

The rod promised essential information for the mine surveyor. Notwithstanding Beyer's reservations about using mineral fragments to augment the attractive power of the dowsing rod, he described a method for determining the underground location of a digger. When dowser and digger both carried iron pieces, the rod above ground might respond to the attraction that developed when the respective pieces came within proximity, indicating that the digger was directly below. Also, in the shafts and galleries, the miners frequently needed to determine whether tunneling should continue in the same direction, or whether the mineral vein had diverted. The method similarly

---

[73] Beyer, *Unterricht*, 13.

[74] "Es schläget die Ruthe auch manchem so leichte, daß er alle Klüfftgen damit findet, und vor Gänge angiebet. Dem Irthum wird aber zum Theil damit abgeholffen, wenn 3. 4. oder mehr Ruthen in einander geflochten und zusammen genommen werden, da denn solche zusammen geflochtene Ruthen nicht auf Klüffte, sondern auf rechte Hauptgänge alleine zu schlagen pflegen." Beyer, *Unterricht*, 12.

[75] "Und wenn zwischen den Ruthengänger und des Marckscheiders Befinden ein Unterschied, muß dieser darinnen mehr auf seine Kunst bauen und trauen...als auf der Ruthengänger ihr Angeben sich verlassen; Treffen einander aber beyde, ist um so viel besser und bewisser sich darauf auch zu verlassen." Beyer, *Unterricht*, 12.

[76] "Dass bey guter Uberlegung und rechten Gebrauch derselben viel Gutes mit der Ruthen auszurichten ist, ist wohl unlaugbahr; Sich aber ganzlich darauf zu grunden und zu bauen, ist niemanden zu rathen, weil sie vielen Verhindernissen und Zufallen unterworffen ist." Beyer, *Unterricht*, 15

involved using a sample of the ore and the dowsing rod. Whether the mineral content of ore increased above or below a given tunnel was also invaluable information for miners. Finally, the surveyor, with the help of a dowser, could determine where to dig a vertical shaft that would intersect a horizontal tunnel below.

Beyer did not make explicit reference to 'Valentine' on vapors, though Beyer, like Rößler and Löwel, clearly believed that metals released mineral vapors in the course of organic generation and decay. When two veins intersected, for example—the 'cross' about which Rößler spoke—the respective *Witterungen* were disturbed, which distorted the dipping motion of the rod.[77] Beyer developed his thoughts on the generation of ore in a chapter entitled, "The Composition and Generation of Veins and Rocks." Beyer began with Creation, the separation of water from earth, and the production of metals on the third day. Just as God propagated plants and animals on the earth's surface by means of 'seed,' so did He invest metals with a generative seed. Furthermore, and now echoing the old Joachimstaler pastor, Mathesius, more clearly, "just as everything above ground eventually reaches perfection, and when the time comes, declines, so the same happens underground; the metals and ore increase, and in the end are consumed and decline, releasing vapors."[78] Beyer, rehearsing standard mining language (*Bergsprache*), like Mathesius again, added that when the miners discovered the putrid remains of mineral ore, they thought they had, "come too late."[79] The Freiberg surveyor described the hollow remains of mineral veins to be like a dead beehive that once contained honey and life.

Beyer did not uncritically adopt these beliefs without awareness of more recent learned theory. When treating of the compass, for example, he mentioned that certain philosophers believed that corpuscles of magnetite raced across the world from the North Pole to the magnetized needle, forcing it to point northward. This idea seemed absurd to Beyer, considering how much interference and space lay between the two extremes.[80] On the subject of metallic generation, Beyer also had

---

[77] "Vermuthlich weil von beyden Gangen jeder seine Witterung oder Ausdunstung hat, und einander irritiren." Beyer, *Unterricht*, 13.

[78] "Und wie alles über der Erden zu seiner Vollkommenheit nach und nach kommt, und wenn es darzu gelanget, endlich wieder abnimmt; also geschicht solches auch über [sic] der Erden, dass die Metalle oder die Erze zunehmen, und endlich durch die Witterung verzehret werden, und wieder abnehmen." Beyer, *Unterricht*, 17.

[79] Beyer, *Unterricht*, 17.

[80] Beyer, *Unterricht*, 33.

occasion to reference both natural philosophy and alchemical theory. The philosophers (*Physicis*) maintained that the inner heat of the earth rendered damp heats and steams (*Dünste*, *Broden*) volatile, bringing them into contact with cooler earths, where they hardened into different sorts of stones, minerals, and gems. The (al)chemists (*Chymicis*), continued Beyer with sarcasm, "want no other material than sulfur and mercury."[81] These principles were the female and male seed, respectively, that mixed with earthly materials in the generation of metals.

Beyer did not reference Latin texts, but rather contemporary German works or translations into German, especially Tobias Beutel's *Arboreto Mathematic*. According to Beutel, the planets and stars also had influence on metallic generation via air and water. The sun guided airs and waters through the crevices and cracks to impart a life-giving power in the generation of gold; the moon did much the same in the birth of silver; and so on.[82] Beyer listed the standard associations between planets, metals, and stones. Mars, for example, ruled over iron and the magnet, but also hyacinth and amethyst. While Rößler, more in line with Thölde and Mathesius, came out against astrology, it would seem that Beyer and Löwel preferred to report the astrological beliefs of fellow miners and dowsers. For more involved treatment, Beyer recommended one "ancient manuscript" by Hans Uttman.[83]

Having sketched the latest in academic and hermetic opinions, Beyer proved himself skeptical of theory in general. Exactly how God imparted growth and decay to minerals, and exactly where the seminal power resided were questions that Beyer left for the quibbling of 'high' scholars (*Hochgelehrten*). He quipped that 'deep' scholars—miners—had more practical concerns: "On this the deep scholars (namely the miners) do not really care. They are worried only about charting the veins and surrounding earths, and how to excavate it all profitably and bring it to a good state. We will leave it to the high scholars, the natural

---

[81] Beyer, *Unterricht*, 18.

[82] "So führeten diese beyden Elementa, Lufft und Wasser, durch solche Klüffte und Eingänge der Sternen Kräffte mit sich hinab in die Erde, und *communicireten* und infundireten dieselbe denen Materiis." Beyer, *Unterricht*, 19.

[83] Beyer, *Unterricht*, 19. Hans Uttman, *Bericht, von denen Ertz-Gebürgen, Streichenderer Gänge, Stöcke, Flöze, Klüffte, Ertze, Berg Arthen und allen Metallen, auch von Schürffen, Seiffenwercken und andern Arthen der Bergwercken* (1601), DHSA, loc. 36070. We know little about this *Oberbergmeister* at Annaberg, though Uttman may have been related to Christoph Uttman, who headed the famous Grünthal liquation works. The manual was printed in 1732.

philosophers and chymists, who are still not in agreement, to figure it all out better."[84]

Beyer exposes an amount of disenchantment with scholarly methodology or rationality in general. He even claimed that critical thought interfered with the functioning of the dowsing rod. In his over fifty years of experience, the simplest and youngest practitioners made the best *Rutengänger*. The more speculation on its operation, by contrast, the less effective were the practitioners.[85] Beyer rehearsed a number of standard questions meant to cast doubt on dowsing practice, but he allowed that answers were not always forthcoming. "What is the reason that the dowsing rod does not dip for all people, or for others only at certain times, or for only a set period, or first in old age, or when one changes or gets married?" Beyer knew a "distinguished man here in Freiberg," for whom the rod never dipped during his first marriage, but dipped during his more recent marriage. For Beyer, these were facts, and their apparent absurdity only proved the futility of attempting an explanation. Dowsing was as great a mystery as the attraction between a lodestone and iron, or between quicksilver and gold, which were considered facts in Beyer's day.[86]

A Side Branch?

This chapter began with Hans Baumgärtel's claim that miner beliefs were merely a side branch in the history of mining science. He considered Augustus Beyer in particular to represent a group of mining officials who circulated among learned men, but who otherwise lacked a university education and awareness of new science. Baumgärtel

[84] "Darum haben sich die Tieff-Gelehrten (nehmlich die Bergleute) wohl nicht sehr bekümmert, sondern ihr Trachten gehet nur dahin, wie die Gänge und Flöze auszurichten, und solche mit Nuz gewonnen und zu gute gemachet werden konnen; als wollen wir es auch denen Hoch-Gelehrten, denen Physicis und Chymicis, noch ferner überlassen, dieses besser auszumachen, die ohne dieß noch nicht einig seyn." Beyer, *Unterricht*, 18.

[85] "Wie ich denn bey meinen uber 50 Jahringen Marckscheider-Dienst mit vielerley Ruthengangern zu then gehabt, und angemercket, dass die einfaltigsten und die mit der Ruthen zu gehen noch unerfahren, die besten sind, denen es am ehesten zugetroffen habe. Wenn sie aber einige Zeit oder Jahre als Ruthenganger sich brauchen und duncken lassen, klug und verstandig zu seyn, auch vermuthlich die Gedancken nicht zusammen halten, oder in solchen Reflexiones machen, haben sie hernachmahls am meisten gefehlet." Beyer, *Unterricht*, 15.

[86] "Also ist in der Natur noch vieles in der Ruthen, gleichwie des Magnets Krafft, Eigenschafft und Liebe zum Eisen, oder auch des Quecksilbers zum Golde noch verborgen." Beyer, *Unterricht*, 11.

concluded that this lack of formal education accounted for Beyer's interest in "mystical and alchemical theories."[87] But that is not a satisfactory explanation, insofar as university or academy training among mining officials, and exposure to new philosophy before the Freiberg Academy, remained the exception rather than the rule, and as the theories to which Baumgärtel was casually referring were long-standing miner beliefs, recorded in mining texts, mining-town sermons, and town chronicles since the earliest printed books. Mining officials such as Löwel, Rößler, and Beyer incorporated prospecting and dowsing knowledge into their *Bergwissenschaft*, as they upheld a new standard of systematization and professionalization. At the vanguard of mining science and practice during an economic rebound in Saxon mining, which included cutting-edge mine surveying, these officials did not imagine a radical break with the accumulated wisdom of centuries, but rather advanced a vernacular mining science.

This is not to suggest that all contemporary mining books were as accommodating of miner traditions: mine surveying manuals in particular, filled with pure mathematics and intended for an elite audience, could just as easily disregard dowsing and miners altogether. Erasmus Reinhold's work of 1574 gave no indication that surveyors ought to consult dowsers, nor did Andreas Böhm's of 1759, to cite just two examples. The link that men like Rößler and Beyer made between surveying and dowsing required defense. We can understand how and why mine surveyors actually employed dowsers only by stepping outside the texts and returning to the mines themselves. We already witnessed surveyors working with dowsers in the salt expedition of 1713/1714, and in the following chapter, we expand and develop our microhistory of dowsing practice into the 1740s.

[87] Baumgärtel, *Vom Bergbüchlein*, 99.

CHAPTER FOUR

TRUE STORIES OF FREIBERG DOWSERS

Superintendent of Mining at Freiberg, Augustus Beyer, once came to the village of Reichenbach looking for a skilled dowser. Beyer was informed that an "old and honest carpenter" lived in the region that could find hidden treasures with a metal ball and some string. Beyer found this man and followed him into the hills, where he hung a small ball on the two middle fingers of his right hand, stretched out his arm, and walked slowly to the mines. As the carpenter approached the mines, the superintendent was amazed to see the ball begin to twirl. The closer the carpenter came, the faster the ball twirled. Beyer wanted to try for himself, so he took the ball and string and tied it to his fingers just like the carpenter had. The man then grabbed Beyer's wrist like a physician checking a pulse and led him slowly to the spot. The ball began to twirl. "I will leave it to the scholars to figure out the cause of this effect," said Beyer. "Be assured, it was no superstition, spell, or deception."[1]

While the account above reads like folklore, in fact it was the testimony of the same mine surveyor we met in Chapter Three. Beyer told of this encounter of around 1720 in the same mine surveying book. From a modern perspective it is paradoxical that one of the highest officers in Saxon mining, Superintendent at Freiberg, would consult a dowser, and we are inclined to assume that the encounter was apocryphal or folklore. But the tales that folklorists collected beginning in the late eighteenth century derived from real events.

After the Thirty Years War, Saxon mining figured more centrally than ever before as the fiscal arm of the state, and bureaucrats and officials sought more reliable knowledge of mineral resources. Mining officials developed a science that legitimized the dowsing rod as the proper instrument of the *Bergverständiger*, or experienced miner, and that distinguished it from superstition and witchcraft. We now go to the Erzgebirge to ask again how this vernacular science played out in the field. Freiberg in particular was at the vanguard of mining science,

---

[1] Beyer, *Unterricht*, 222.

and that included the patronage and evaluation of dowsers. 'True stories' of Freiberg dowsers show that the formulation of mining science was the product of real struggle in the field, much as it involved new laboratory practices. Officials and chemists like Johann Friedrich Henckel were practical-minded and accepted the possibility of dowsing, but they were also skeptical, bringing dowsing and dowsers into a more critical spotlight than at the Dübener Moor in 1713/14. Again, this is uncharted territory in the historiography of mining and geology, which has neglected the widespread employment of dowsers by the Saxon mining jurisdictions, including the central office at Freiberg (*Oberbergamt*). The phenomenon offers a unique insight into the values and beliefs of miners and mining officials, and their effort to define and authorize mining knowledge just decades prior to the founding of the Freiberg Mining Academy (1765).

Officials in the Erzgebirge addressed the patronage of dowsers in various papers, among them the regular reports from the field on conditions at the mines (*Grubenberichten*), and correspondence between the mining jurisdictions and with the Dresden mining administration concerning hired dowsers. This chapter analyzes a sample of such archival material spanning the first half of the eighteenth century. We select a packet of mine reports for the Saxon jurisdiction of Oberwiesenthal high in the Erzgebirge, and we analyze a number of letters and contracts between Freiberg and Dresden officials concerning individual dowsers. This material will also help us develop a profile of the hired dowser. While idiosyncrasy was his trademark, the legitimate practitioner exhibited characteristic traits and behaviors. These included humility, deference to officials, and appeal to divine grace. He demonstrated a tacit knowledge of dowsing and had (preferably formal) mining experience. He was a *Bergverständiger*, or knower of the mines and mountain. The epistemological status of the rod was tied to the ability of the practitioner to demonstrate that identity. Immodest claims and boasting, utterance of magical words, use of the rod as an instrument of sorcery, and inexperience at the mines, all suggested incompetence or superstition to the mining jurisdictions. Historians and sociologists of science have emphasized the significance of character and social credibility in establishing truth-claims largely in terms of how gentlemanly culture verified new science.[2] Here, we witness a

[2] Some examples include Martin J.S. Rudwick, *The Great Devonian Controversy: the Shaping of Scientific Knowledge among Gentlemanly Specialists* (Chicago: University

similar phenomenon among mining officials, but toward the legitimization of a different sort of knowledge.

Mining Reports

One of the most important steps in organized mining was the gathering of a group of shareholders (*Gewerkschaft*), the men who purchased a set number of shares (*Kuxe*) and effectively subsidized the equipment, machinery, and cost of labor. According to the mine ordinances, the mine master authorized formal digging after mineral earth had been discovered and reported. The on-site mining officials—especially shift managers (*Schichtmeister*) and account managers (*Lehnträger*)—kept a close watch over all activities to ensure that claims proceeded legally, and reported regularly about new prospects. Their reports expose the day-to-day concerns of local officials as they sorted through hundreds of pronouncements coming from the field, and mediated information to their superiors at the jurisdictional office, the central office at Freiberg, and the mining administration at Dresden. Independent prospectors in the Erzgebirge happened upon silver- and copper-bearing earth, iron deposits, tin, bismuth, and cobalt, among other minerals. Accounts of these finds were detailed and professional, and because they might serve to attract more attention and investment to a given area, they were decidedly optimistic, if not exaggerated. The authors stressed how vigorous mining had become in their region, and how numerous the finds and high the prospects, often implying that a great boom in production was impending.

A typical report covered three main issues: it began with a brief history of the local mines, including the negative effects of prior wars or flooding, and when higher intensity prospecting and digging had resumed; it listed actual prospects, claims, and conditions, including the direction, depth, and consistency of known mineral veins, and the results of any available assays; and finally, the reports reviewed the status of the shareholding group, including how complete it was, whether shares remained unsold, and whether it had hired labor to begin work. It is clear that the 1713 salt expedition (Chapter One) was

of Chicago Press, 1985), Mario Biagioli, *Galileo, Courtier: The Practice of Science in the Age of Absolutism* (Chicago: University of Chicago Press, 1993), and Steven Shapin, *A Social History of Truth: Civility and Science in Seventeenth-Century England* (Chicago: The University of Chicago Press, 1994).

not exceptional, as dowsers numbered among the men who provided information on conditions and prospects at Oberwiesenthal. A trusted dowser could influence the decision-making of officials and investors.

We might expect the pronouncements of dowsers to resemble the far-reaching or irrational productions of common soothsayers and other popular magicians, but here again we find only the sober language of mining officials as they recorded matter-of-fact information provided by dowsers in the field. The authors presumed themselves very capable of distinguishing trustworthy and unlikely statements, and not all dowsers were worthy of their time: the officials showed a preference for well-worn or experienced men whose claims had been (at least in part) substantiated by actual digs; men from the region with a proven record of honest dealings, especially with other mining jurisdictions like Freiberg. As in the Dübener Moor, the authors here dropped the derogatory prefix 'Wünschel-' ('wishing,' or 'dowsing'), preferring more everyday language as 'Rutengänger,' 'Rutengeher,' or 'Rute' (rod). In the some 50 reports spanning the 1730s for Oberwiesenthal, almost 200 pages of material, I find only one usage of the more derogatory language.[3] By contrast, the authors introduce true dowsers with such honorific descriptors as 'valued' (*bewehrten Ruthengänger*), or 'well-known' (*beruhmten Ruthengeher*), a strong indication of the respect afforded the legitimate dowsers or *Bergverständiger* more broadly.

This does not mean that Oberwiesenthal and other Saxon jurisdictions were so naïve as to believe all dowsers without discrimination. As we shall see, some dowsers came under suspicion, and by studying such cases, we can clarify the distinction contemporaries made between legitimate and illegitimate dowsing. We might expect that to be made empirically through experimental examination, but an exclusive reliance on scientific procedure was not common in German mining before the nineteenth century. Rather, the hired dowser was an experienced miner, and he demonstrated a certain humble simplicity and piety, whereas the false dowser failed to demonstrate these qualities. Some level of empirical proof was always expected in trial digs, but the results were subject to various interpretations, and nobody relied exclusively on them to evaluate a dowser's work.

---

[3] BAF, Loc. 40014, Nr. 74, Fol. 160.

An extract from November 1, 1732, can serve to introduce the language of a typical mining report, and demonstrate how well integrated dowsing was in the business. The Blessed Baker complex near the town of Fichtelsberg consisted of various mines, among them Divine Help. At this particular location, miners tracked any number of veins and deposits. The following selection concerns a vein the miners called Dietrich von Bären, likely named after its founder. It addresses the work of dowsers, and reads as follows:

> We expect to find and work rich earth beside Divine Help, as two well-known dowsers, neither of whom had seen the other, nor had any knowledge of the other's claims, both, after walking with their rods, believe that...Dietrich von Bären vein will cross [other] very strong and rich veins at 45 Lachter [90 meters], then again after 2.5 Lachter [5 meters]... and yet again after 2.5 Lachter, then 2 Lachter after this [4 meters], and especially in 4 to 5 Lachter [8 to 10 meters] thereafter... The same dowsers have also claimed that 45 Lachter from the main shaft, 2 Lachter deep, there is fahlore. In general these works enjoy great prospects... with reports of fine mineral earth breaking, such as silver-bearing Glanz, pyrite [Kies], black, yellow, and brown earth... Without doubt, anybody who, with God's blessing, tries his luck here will not have waged his money in vain.[4]

The first sentence demonstrates that officials were careful in their dealings with dowsers. The fact that two practitioners, entirely unknown to each other, would issue such comparable statements about prospects at Dietrich von Bären gave the claim more validity and secured it a place in these official papers. That the author describes the dowsers as "well-known" further emphasizes that this was not the production of swindlers or magicians, but rather the informed opinion of experienced miners and dowsers.

Some of the language that follows might appear vague or general, but it should be understood in context. Beyond specifying fahlore, the dowsers do not determine the mineral constitution of the earth they detect beyond describing it as 'strong and rich.' When the author mentions the earth that miners were actively discovering, he specifies its color—'black' and 'brown.' It is important to remember that chemical mineralogy, which would develop in Freiberg, was in its infancy, and that crystallography was nonexistent, so that more exact determinations were impossible. Both officials and dowsers, in Saxony and

[4] BAF, Loc. 40014, Nr. 74, Fol. 15b–16.

abroad, used terms such as 'strong,' 'reddish,' 'golden,' 'shiny,' 'brittle' [*spätig*], or 'metallic' [*gediegen*] to describe mineral earth before an assay confirmed any metallic content. Miners, mineralogists, and dowsers alike all distinguished rocks and minerals by taste, color, hardness, smoothness, and other observable characteristics. When they specified pyrite, fahlore, cobalt, or another mineral, we should also remember that mineralogists today distinguish any number of pyrite rocks, fahlores, and cobalts, and so we cannot expect to know with certainly which substances our actors were describing. The author, using good miner language (*Bergsprache*), claims that these ores are "breaking," by which he means that the miners were accessing the veins and lifting rock.

The dowsers' measurements too might seem haphazard, as if they deliberately addressed large tracts of land under the assumption that diggers would validate one or another of the claims. But the officials did not share such skepticism. Dowsers stressed that their art was delicate and dependent on various factors, ranging from the mental condition of the practitioner to unknown geological formations, and few expected a given pronouncement to correspond with a given find to the Lachter. It was widely assumed that the rod dipped to any number of underground objects—water sources, fossils, caves, springs, mineral earths—so the dowser *should* exercise restraint in his claims, otherwise he might attract suspicion. Both officials and dowsers would expect that prospective digs revealed something of interest or value, but all parties were content to dig without absolute a priori certainty.

In defense of the dowsers of Divine Help, finally, the claim that fahlore lay two Lachter (4 meters) deep was not beyond easy verification by a team of diggers in dry conditions. The skeptic might rather assume that dowsers—swindlers, one and all—would report mineral earth at greater depth. The last comment that the purchasing of shares would "with God's blessing" yield good returns underscores the casual religious sanction behind mining and capital investment that was common, notwithstanding certain theological misgivings.

Reports like this recorded the many predictions of dowsers as a form of self-promotion or advertisement. "All manner of dowsers" had agreed about prospects at Joyous Hope, boasted a report of February, 1735.[5] "All credible dowsers" were certain of good prospects at

[5] BAF, Loc. 40014, Nr. 74, Fol. 81.

Fichtelsberg in February of 1736.[6] The author of the March 22, 1732 report for the St. Michaels Mine near Unterwiesenthal was proud to declare that "various dowsers" had found the area surrounding the main dig to be rich in mineral earth. He then described actual conditions and prospects, including the following: "Already at the 9th Lachter [18 meters] a vein has been crossed containing all manner of mineral earth, such as yellow and brown, and within 4 to 5 Lachter [8–10 meters] two main silver veins should be found, which are, according to the strong dip of the rod, very noble."[7] Again, we see that earth was described mainly by color, before learning that the quality of the earth correlated to the strength with which the rod dipped. This becomes a trope in these reports, as they sought to generate more excitement: at St. Michael's mine, May, 1735, the "strong dip of the rod" revealed mineral earth to be rich and noble; and "the powerful dip of the rod" indicated rich silver-bearing earth.[8] In 1737, again, miners expected to unearth silver, because of the "strong dip of the rod."[9]

A further report for August 12, 1732 introduces more language and helps us begin focusing on particular well-known dowsers. At Deep Rosencranz, the local miller, Michael Rosencranz, an "old and respected man," had first reported the presence of rich mineral earth, "[later confirmed as] veins and tributary or feeder branches [*Trümmererz*] containing yellow and brown earth, and also fine pyrite with marcasite and blue seams [*Flösse*], assayed at 4 to 5 Lot of silver."[10] The preference among mining officials for local knowledge extended not only to dowsers, but also to all villagers on the mountain, such as this miller, who were assumed to possess credible knowledge of the earth. The bluish earth may have been a species of cobalt, and 4 or 5 Lot (~60–75 g) marked the upper limit of the silver content determined by assay for the majority of ores. More modest 1 or 2 Lot determinations were frequently noted.

The Deep Rosencranz report then continued with the "respected dowsers" who, after "a prompt and thorough walk-about" [*schleunigen und beständigen Umbtrieb*] had unanimously pronounced the area rich in ore. Most notably, a highly respected dowser had walked the region.

[6] BAF, Loc. 40014, Nr. 74, Fol. 118b.
[7] BAF, Loc. 40014, Nr. 74, Fol. 34b–35.
[8] BAF, Loc. 40014, Nr. 74, Fol. 65b–66.
[9] BAF, Loc. 40014, Nr. 74, Fol. 124b, 125, 127b
[10] BAF, Loc. 40014, Nr. 74, Fol. 47. One Lot ~ 15g. One Mark ~ 16 Lot.

He was, "the sworn royal dowser from Freiberg, Michael Christian Knorr," whose four pronouncements the author listed.[11] Knorr's title has great significance: he was employed by the Central Mining Office (*Oberbergamt*) at Freiberg, and was, accordingly, more directly at the service of the Elector in Dresden. This explains the loosely applied designation, 'royal.' He had already emerged on June 20, 1732, as the, "royal dowser Herr Knorr from Freiberg."[12] Knorr would have enjoyed considerably more credibility than other dowsers, introduced only anonymously in these reports, who did not have such high qualifications. Saxon mining jurisdictions often turned to Freiberg, the central jurisdiction, for reliable and tested dowsers such as Knorr. We shall have additional indications that at Freiberg dowsers were sworn men, or that they took formal oaths and handshakes like windlass men, mine surveyors, shift managers, and other workers and officials of various rank.

Among his claims, Knorr believed that diggers would find some ten rich veins 24 Lachter (48 meters) from the main dig at Rosencrantz mine, but in a mere 4 Lachter (8 meters) they would discover a vein containing "white-gold" earth, especially Glanz [*Silberglanz*].[13] Knorr appears in yet another report for Hauenstein of June 28, 1732, as the "famous dowser from Freiberg." Here, among many other claims, Knorr detected that the common rock that lay beside the St. Wolffgang vein would transform or 'purify' [*veredeln*] into "noble reddish-gold earth and cobalt."[14] It was widely assumed, based on biblical and alchemical sources that minerals grew organically like plants and animals, and therefore miners hoped to track mineral veins to their better developed, or more purified and noble forms.

But another name dominates the reports of March, 1732. Johann Gabriel Schreiber from Abertham, another important mining town in the region, was a "very famous dowser" who, the author explained, had been working for years in a number of distinguished mining towns. Schreiber was in great demand since, "everyone knows that whatever he has claimed with his rod, at various mines in neighboring mining towns, especially at Johanngeorgenstadt and St. Joachimstal, has been discovered most accurately and exactly." He was now in the Pochwerk

---

[11] BAF, Loc. 40014, Nr. 74, Fol. 47b.
[12] BAF, Loc. 40014, Nr. 74, Fol. 21.
[13] BAF, Loc. 40014, Nr. 74, Fol. 47b. Silberglanz=Acanthite, a silver sulphide.
[14] BAF, Loc. 40014, Nr.74, Fol. 25.

region (Pochwerk=ore crushing machine) and had performed quite satisfactorily at the Joyous Hope Mine. Schreiber had claimed to locate alluvial deposits containing "silver-bearing Glanz [*Silberglanz*]," and some Lachter beyond that, another "rich vein." He specified at least two more silver-bearing veins, and diggers had confirmed at least one of these, the depth of which (1.5 Lachter or 3 meters) Schreiber had predicted precisely. Among other pronouncements, Schreiber detected a layered mineral rock (*Stockerz*).[15] Because Schreiber's claims were proving true, the miners made additional prospective digs and found "reddish-gold earth" at another spot the dowser had indicated.[16] The author does not explain how exactly Schreiber's statements corresponded to these new finds. The dowser had detected some sort of rich earth, and some sort was found.

Schreiber's reputation preceded him, as other reports for nearby mines included the same summary of his history and credibility that the report above had introduced. At Fichtelsberg, before describing the dowser's work, the officials repeated the personal information almost verbatim, adding only that Schreiber secured his "livelihood" (*Profession*) from dowsing.[17] Many dowsers merely supplemented their other wages at the mines, but this statement suggests that Schreiber had specialized in dowsing. At Young Tobias vein, the report included the now standard introduction on Schreiber, adding that his high performance had been, "reliably reported."[18] At this region, miners had discovered "reddish-gold" earth where the dowser had indicated. One of his claims read as follows: "In 21 and ¾ Lachter [43 meters] from the central dig [*Feld Orthe*], there will be a main vein, from which, near Divine Help, metallic [*gediegen*] earth will break, and on which major works will be built; and at the feeder veins [*Trümmer*] by Divine Help, mineral ore will also break, beyond any doubt."[19]

In all, the officials in Oberwiesenthal alone referred directly or indirectly to Schreiber's work at least 10 times for the months of March and April, 1732, amounting to well over 50 individual predictions or claims. We can assume he worked as diligently in any number of regional towns, producing many hundreds of claims. His statements, as all

[15] See Grimm and Grimm, *Deutsches Wörterbuch*, s.v. 'Stockerz.'
[16] BAF, Loc. 40014, Nr. 74, Fol. 17b–18b
[17] BAF, Loc. 40014, Nr. 74, Fol. 40b.
[18] BAF, Loc. 40014, Nr. 74, Fol. 49.
[19] BAF, Loc. 40014, Nr. 74, Fol. 49.

others recorded by the shaft and account managers, illustrate just how well versed in the language of mining in general, and in local conditions at Oberwiesenthal in particular, these dowsers were. There were swindlers and traveling magicians to be sure, but only a true miner could demonstrate such experience and number among the publicly hired dowsers.

Knorr and Schreiber, however admired and sought-after, were not as famous as Tobias Häusler, who, after decades in mining, rose to prominence in the late 1730s. Earliest mention dates to February of 1738, where we hear of the claims of a "famous dowser from Röthenbach."[20] By late March, we have a more complete picture of this man whom Freiberg would investigate and expose as a fraud. But before treating Häusler and his work, and studying how Freiberg evaluated Häusler, we might introduce dowsing at Freiberg more broadly, site of the Central Mining Office (*Oberbergamt*) and future home of the Mining Academy.

The Patronage of Freiberg Dowsers

The head overseer of Saxon mining, Abraham Schönberg, did not discuss dowsing in his mining book of 1693, which historians of mining rightly view as highly progressive. He merely relegated it to a glossary of terms and phrases. Yet, manuscript sources complicate the argument that officials like Schönberg contributed only to rationalization or modernization in mining. A letter of 1709 to the mining administration at Dresden (*Bergrat*) concerning local dowsers suggests that Schönberg's printed work did not address his interest in promoting dowsing. Recognizing a succession of dowsers in the region, he and two other high officials ensured that those employed by the Freiberg Mining Office (*ex officio*) were paid in the same manner as other miners and recognized for their work.[21] Schönberg sought to incorporate and oversee dowsers more formally in the field, the very year the

[20] BAF, Loc. 40014, Nr. 74, Fol. 106

[21] "...wenn die Ruthengänger von denen Berg Beamten zur Ausgehung ex Officio erfordert werden, ihnen das Lohn auf denen Stölln einen Weg wie den andern verschrieben werde; Also hingegen daferne...von denen Gewercken die Ruthengänger vor ihre Mühe absonderl. vergnüget werden sollen." BAF, Loc. 2092, Fol. 2b. In August Friedrich Wappler, "Alte sächsische Wünschelruten-Geschichten" in *Mitteilungen des Freiberger Altertumsvereins* 43 (1907), 66.

Elector renewed the fund that Schönberg had originally proposed to support teaching, and which included support for digging expeditions (*Schürfen*). By using the term "Ruthengänger," moreover, he too demonstrated everyday familiarity with the 'rod' and downplayed its otherwise questionable associations (*Wünschel*rute).

The distribution of sources suggests that formal patronage at Freiberg began in the later seventeenth century, but that the high point was the mid-eighteenth century. We see in correspondence of the Freiberg Office and that of other Saxon jurisdictions (*Bergämter*), beginning in 1713 and continuing into the 1740s, that the Office employed dowsers from at least the time of Schönberg to find and determine the properties of mineral veins, to locate the boundary markers between mines, and to discover hidden coins, jewels, and other valuables. We develop our profile of the hired dowser considerably as we work through this correspondence.

Augustus Beyer mentioned a curious practitioner in his mine surveying book who responded so violently to mineral vapors, with or without his rod, that people called him laconically 'Shaking Hans' (*Rüttel-Hans*).[22] He may have been the same sickly fellow about whom Johann Voigt wrote in 1713. This case illustrates the open-minded skepticism of the officials and begins our character profile of the Freiberg dowser: the true practitioner was a hearty miner, not a delicate scholar. Voigt—the *Oberbergamtsverwalter* who had authorized the salt expedition of 1713, we might recall—described a certain Hans Wolff, born to an oil miner in Zethau, who had been mining in the region for nine years. Now thirty-five, the dowser was, according to Voigt, a simple man of weak constitution (*sehr einfältig und von kränklicher Leibes-Constitution*) who wasted his money on books by the alchemical thinkers, Theophrastus Paracelsus and Basil Valentine (Thölde), and the magus, Cornelius Agrippa.[23] In Thölde's *Report on Mining* of 1600 (and the museum curator and Bergrat at Freiberg, August Friedrich Wappler, later suspected that Wolff possessed the *Last Testament*, which contained the *Report*[24]), Wolff would have found a complex theory of mineral vapors and their effects on the dowsing rod. That Voigt, who was skeptical of the dowser, added that Wolff was

[22] Beyer, *Unterricht*, 221.
[23] Wappler, "Wünschelruten-Geschichten," 56.
[24] Wappler, "Wünschelruten-Geschichten," 57.

still unmarried suggests that the ideal dowser was an established and employed man, master of both home and nature.

Wolff, believing that the rod dipped only for individuals graced by God, and claiming to work a dig that would reach mineral earth at 3.5 Lachter (7 meters) deep, now desired to employ his gift at Freiberg. His method involved holding out his right arm, concentrating on the ground, and pacing the field slowly. When he approached an underground vein, his arm and body shook so violently that he became exhausted. Voigt was concerned for Wolff's health.[25] Having located a vein, he then used a large onion-shaped rod to determine more exact properties. He had known from a young age that the "rod dipped for him." It dipped so well, in fact, that it struck Wolff violently in the chest until he could barely hold on.[26]

Voigt remained skeptical. Wolff responded markedly and immediately upon grasping his rod, and his reaction increased in strength the closer he got to a vein, unlike other *Rutengänger*, for whom the rod dipped mostly at the veins themselves. Voigt questioned Wolff on this uncontrollable shaking: "How could he dig and work mineral veins when they gave him such strong movements in his body?" Wolff answered that the movements stopped when he remained indifferent as to whether he would strike a rich vein.[27] One must not interfere in the workings of the rod, he explained, and risk compromising its natural attraction to mineral ore; nor should one test and compel the rod by searching for hidden coins, he added, which was a common procedure. What he was describing was a subtle interaction between mind, body, and nature: Wolff stressed the importance of mental control, but he also believed in a physical, albeit invisible, connection between mineral ore and the dowsing rod. That influence, like magnetism, grew stronger the closer his rod came to the source.

Voigt then tested the presumed dowser by bringing him to a region under which stretched earth that had not yet been worked. He asked Wolff to determine the depth at which it became mineral. The dowser

---

25 Wappler, "Wünschelruten-Geschichten," 55.

26 "...vermehrete sich das Schlagen der Ruthe...daß man die Schläge, so sie Ihm gegen das Hertze gab, gar weit hören kunte, und vermochte Er dieselbe nicht zu erhalten." Wappler, "Wünschelruten-Geschichten," 56.

27 "...dass wenn Er nicht seine Gedancken auf die Findung eines edlen Ganges gerichtet, so thäte es Ihm nichts und könnte darüber hingehen, und auch darauf arbeiten." Wappler, "Wünschelruten-Geschichten," 56.

preceded alone a short stretch before raising his rod. Voigt was confused when Wolff spit at the instrument three times before reporting its findings. It is unclear whether the dowser, who, according to Voigt, seemed ashamed at the moment, incorporated his phlegm into his practice or merely rid himself of it.[28]

Other cases echo the mining reports we studied above by demonstrating how renowned Freiberg was for its dowsers. In October of 1728, the Stollberg City Council wrote the Central Mining Office, requesting that it send a "sworn" dowser to help settle a boundary dispute.[29] Here is another indication that the dowser might actually have performed an oath or received license, otherwise common in the trades, and suggests again how dowsers acquired a legitimate status. Stollberg may also have had a tradition of using dowsers to locate boundary-markers, since a similar case occurred in 1703, when "Ruthengänger" Christian Vogel successfully discovered and unearthed old markers (*Reinsteine*). The opposing party claimed that his work was unlawful and suspicious, but Vogel had been formally contracted since 1700 to settle land disputes between Saxony and Bohemia.[30] In the 1728 case, Stollberg had already corresponded with the Schneeberg and Marienberg mining offices, but was informed that no dowser was willing or available, and that it might appeal rather to Freiberg. In their letter to Freiberg, the Stollberg officials gave an exact date for the arrival of a dowser, specified his contact, and proposed to pay him after the

---

28 "So wandte Er sich von mire weg, nahm seine Ruthe aufrecht, und wie ich genau observirte, spye Er dieselbe ganz geschwinde dreymahl an…Das Anspeyen der Ruthe wollte Er auf beschehene Vorhaltung nicht gestehen, sondern sagte, daß er sonst also ausgespyen." Wappler, "Wünschelruten-Geschichten," 56–7.

29 "…einen geschwohrnen Ruthengänger welcher mit seiner Ruthe die streidtigen Reinungen [Rainsteine] aufsuchen kan." Wappler, "Wünschelruten-Geschichten," 59.

30 "In streitigen Reinungs-Sachen A.S. contra H. B. haben Potentiss. Rex noster & Elect. Ao 1703 d. 31 August an den Amtmann zu Stollberg allergnädigst also rescribiret: Wir können geschehen lassen, dass, wenn die Vereinung nach Innhalt nach unserer vom 8 May a.c. ergangenen Resolution vorgenommen wird, der Ruthengänger Christian Vogel darzu gezogen werde, und dadurch den Modum, durch die Ruthe Reinungen auszugehen, allergnädigst approbirt, wobey zu verwundern, dass nach der hierbey gehaltenen Amts-Registratur *ernannter Ruthengänger die alten verfallenen Reinsteine richtig angegeben, und sie beym Nachgraben mit ihren Zeugen richtig befunden worden*. Ob auch gleich hernach diese Besichtigung von den Gegnern als unerlaubt und superstitiös gehalten worden; so haben doch die Herren Schöppen zu Leipzig auf das vorhergegangene Compromiss und dabey gebrauchte Eidesleistung gesprochen. Ja es ist den 13ten Sept. 1700, eben dieser Vogel zu Ausgehung streitiger Gränzen zwischen Böhmen und Sachsen ordentlich vorhero verpflichtet worden." In Feudiviro, *Gebrauch der Berg- und Wünschel-Ruthe* (Leipzig, 1763), 10, my emphasis.

job. But Freiberg echoed the other offices by claiming to have no dowser willing to take the job.[31]

It is not clear why Freiberg could not recommend a name, why this particular job did not appeal to any dowsers, or why Freiberg did not share its local talent. That Freiberg was an obvious place to turn is clear in other correspondence. When Dresden learned in the winter of 1713 of supposed jewels buried near the town of Ziegelscheune, just up the river Elbe, it requested "a dowser from Freiberg" to make a more thorough investigation. Expressing concern that the circumstances were "suspicious" and demanded utmost secrecy, the rod appeared here as a reliable forensic tool, and the Mining Office the proper place to turn.[32] In 1735, it was again a "dowser from Freiberg" named Weigolt to whom Dresden appealed for information on the treasure hunting activities of a late official (*Amtsverwalter*) named Pflicke in the village of Nossen, just north of Freiberg. A master mason had named Weigolt to authorities, and Dresden accordingly ordered the man to report to the famous Freiberg Bergrat and chemist, Johann Friedrich Henckel, for further questioning. Henckel was quite familiar with dowsing: in a text on the similarities between the vegetable and mineral kingdoms, the skeptical chemist explained that coming to Freiberg brought him in contact with dowsers, though he preferred to withhold judgment as to the veracity of the practice.[33] In correspondence of 1738, moreover, an Altenburg landowner recounted for Henckel experiences with a "well-liked" dowser (*beliebig Ruthengänger*) who claimed to find coal.[34] In the case

[31] "So leidt thut es uns hingegen, daß wir ebenfalls aus Mangel dergleichen, wie selbiger zu der verlangten Aufsuchung der streitigen Bereinung erfordert wirdt, solches zu thun nicht vermögende." Wappler, "Wünschelruten-Geschichten," 59.

[32] "Ob nur Wohl die angegebenen Umstände gar verdächtig; so sind wir doch entschloßen diese Sache etwas genauer untersuchen zu laßen und begehren dannenhero gndst, ihr wollet zu solchem Ende einen Ruthen Gänger von Freyberg alsofort anher verschreiben, und selbigen die angegebenen u. daherum gelagenen Gegend einschlagen laßen." DHSA, Loc. 7291, Nr. 2, Fol. 20.

[33] "Ich hatte mir vorgenommen, etwas von der *Wünschelruthe* hier beyzufügen, woraus eine ziemliche Wahrscheinlichkeit vor die Verwandschaft besagter beyden Reiche hervorleuchten möchte, weil ich aber nach meiner wenigen Erfahrung und Erkundigung, so ich seit meinem Aufenthalt in unserer geliebten Bergstadt Freyberg einzuziehen nicht vergessen, grose Ungewißheiten und Dunkelheiten dabey finde, so will ich davon lieber schweigen, und dasjenige, so ich etwa sonsten noch zu diesem meinem Vortrag hätte mit beybringen können, bis zu anderer Gelegenheit versparen." Henckel, *Flora Saturnizans*, 562, my emphasis.

[34] J.C. Lorenz to J.F. Henckel, July 26, 1738, in *Mineralogische, Chemische und Alchymische Briefe von reisenden und andern Gelehrten an den ehemaligen Chursächsischen Bergrath J.F. Henkel*, Band II (Dresden, 1794), 333.

of the treasure-hunting official named Pflicke, Henckel learned that the deceased had apparently employed the dowser Weigolt around 1700 to clear some debris or rubble (*Schutt*), and had also asked him "whether he could use the rod" (in addition to his shoveling work). After clearing the grounds, Weigolt proceeded to dowse the area for Pflicke.[35]

Whether Weigolt discovered anything at this site was not stated, though the Henckel report continued with another commission for Pflicke and an unnamed "eminent gentleman." The digger and dowser Weigolt helped these men clear a shaft or mine opening covered with a large stone that ultimately revealed "3 or 4 middle-sized spoons…in a circle." After this strange discovery, according to Weigolt, he was released. He knew nothing more about the treasures or activities of the late Pflicke. Weigolt also informed Henckel of treasure rumored to have existed in a local monastery, of which the dowser was either skeptical or had disproved.[36] The papers show that Weigolt associated with a number of Nossen commoners and artisans aware of local treasure hunting, including a farmer, legal assistant, forger, mason, and postmaster. As Bergrat, Henckel learned of such popular treasure hunting and dowsing.

Another appeal to Freiberg occurred in 1736, when a certain Christian Baumann of Schneeberg found reason to believe that coins and jewels were buried near his house. The head of the Dresden Administration (*Bergpräsident*), Johann Christoph von Hennicke, a close associate of Heinrich von Brühl, a major advisor to the Elector, allowed the dig to proceed at Baumann's own expense. Von Hennicke asked the head tithe collector (*Oberzehntner*) in Dresden to employ two Freiberg *Ruthengänger*. Hennicke wished only that the party use no "superstitious or otherwise suspicious method" and send samples and reports to Dresden.[37] The tithe officer proceeded to write the Freiberg Office requesting a "skilled and informed" dowser. The Office

[35] "…ob er mit der Ruthen auch gehen könte? Da er nun solches bejahet, so wäre er mit der Ruthen gegangen, und hätte auch dieselbe an dem Orthe, wo der Schutt abgeräumet worden, geschlagen." DHSA Loc. 10690, Nr. 47, Fol. 30.

[36] "…daß er vielmahls auf Befragen derer Leuthe zur Antwort gegeben, wie er auch dabey gewesen wäre, ob gleich in der That solches sich nicht also befände." DHSA, Loc. 10690, Nr. 47, Fol. 31b.

[37] "…daß keine abergläubische, noch andere verdächtige Mittel dabey gebrauchet werden, alldieweilln auch besagter Baumann anzuführen weiß, daß öffters Geld von ihm und den seinigen gefunden worden, so habt ihr hiervon etwas einzuschicken, auch wie viel das gefundene gewesen, zugleich mit einzuberichten." Wappler, "Wünschelruten-Geschichten," 60.

sent a certain tunnel manager (*Steiger*) known for his dowsing, Caspar Fischer from Unexpected Luck mine at Kleinwaltersdorf, just outside of town, and granted him a respectable twelve groschen daily for the job.[38] The digger still earned some 20 groschen weekly by century's end, and the regular tunnel manager some 30 (an *Obersteiger*, between 50 and 70).[39] Dowsing supplemented Fischer's more institutionalized position as foreman. In the contract between the local mine manager and Fischer, he is referred to simply as a "Ruthengänger." Officials and bureaucrats like Voigt and Hennicke were cautious and skeptical, but they were willing to employ a trusty Freiberg dowser.

Bergpräsident Hennicke appealed to Freiberg again in 1747, after Johann Friedrich Junghans from Rahnis, near Erfurt, sent an intriguing letter about treasure in his cellar. A Freiberg dowser again exposed the falsity of a popular tale. Junghans explained that one night, fourteen days before Christmas, he descended into his cellar to find something to drink. There, he saw a chest or case, about one and one-half yards long, filled with gold, and upon which lay a golden soupspoon. He stared in amazement for fifteen minutes before deciding to procure a metal object to throw and prove the existence of the treasure; but it was gone when he returned, leaving only a deep pit, which Junghans filled with dirt and stones. News of these events spread quickly, and a Jesuit from Erfurt supposedly volunteered to cover the cost of digging for a portion of the extracted wealth, if Junghans could procure formal authorization from Dresden. We unfortunately learn nothing of this inquisitive cleric. The men (or Junghans alone?) were now asking Hennicke for permission and protection in this undertaking.[40]

Rather than send a letter of authorization, Hennicke dispatched two men from Freiberg, one dowser named Dietrich and one digger. The dowsers did not perform their own digging, but directed the work of the hackers. When the men and their official contact, Johann Schußman, Amtman to Ziegenruck, arrived at Junghans's house, his daughter informed them that her father was gone. They proceeded nonetheless to search the cellar with the rod, finding nothing. Dietrich did not suppose that even "one penny" lay in that cellar.[41] The party waited in vain

[38] Wappler, "Wünschelruten-Geschichten," 61.
[39] Baumgärtel, *Bergbau und Absolutismus*, 45.
[40] DHSA, Loc. 25192, Fol. 1–2.
[41] "Er sich nicht getrauete, einen Heller darinnen zu finden." DHSA, Loc. 25192, Fol. 10b.

for Junghans to return. On his return trip, however, Schußman encountered none other than Junghans himself, who claimed he was too exhausted from the trip to attend the men on a second examination. The following day, the party reassembled at Junghans's home, and he directed the dowser to more exact spots. After the cellar was almost entirely dug up, the men had still found nothing. At this point in the report, Schußman concluded that Junghans, "spoke no true word, ruined simple people, associated with swindlers," and was involved in a related scheme in 1721.[42] Having heard of these events, Hennicke sentenced Junghans to 14 days in prison.

The trusty Freiberg dowser who exposed the deception was Christoph Dietrich, whom Hennicke had already employed at least a decade earlier. According to a letter of the Mining Office of January 1733, he and his companion, Johann Christian Kuxe, both "miners" well known to the local mine manager, also worked as dowsers. We know from later documents that the men had been in mining for sixteen and fourteen years, respectively. The Office resolved to grant a modest six groschen weekly to the dowsers to work the New Deep Prince mine. That would have complemented their mining salaries. Dietrich in particular was to reside, "in town, so that he be more readily at hand."[43] Henckel, with the head overseer of Saxon mining, Adam Friedrich von Ponikau, and the mine manager, Hans Carl von Kirchbach, among other high Office members, signed the document. Two years later, Hennicke requested that the Office continue paying Dietrich's wife his salary while he was in Poland, where the Elector and now King of Poland evidently found use for Dietrich.[44]

Other correspondence involving Dietrich shows how dowsing was a precarious job at the Freiberg mines, and how dowsers constructed an image of themselves as true 'knowers of mining,' or *Bergverständige*. Dietrich and Kuxe defended their practice in a November 1738 letter to the Mining Office. The men first confirmed their legitimacy, recalling having been sworn by the Office as dowsers with a handshake (*den Handschlag als Ruthen Gänger abgegeben*), and claiming to remain "obedient" and "honorable" miners. They proceeded to complain that certain individuals of little mining experience (*Bergwerksohnerfahrene*)

---

[42] DHSA, Loc. 25192, Fol. 12.

[43] "…in die Stadt, umb desto eher an Handen zu seyn." Wappler, "Wünschelruten-Geschichten," 65.

[44] Wappler, "Wünschelruten-Geschichten," 67.

made unfounded claims and stole their business. "Were they miners, it would be tolerable," said the dowsers, but the region was riddled with questionable characters from all walks of life. Dietrich and Kuxe provided a few examples: there was a certain craftsman from whom "everybody has to hear a lot," and who managed to ruin upstanding miners;[45] there was a carpenter from Hausdorf, near Frankenberg, who seemed to be everywhere, but produced results nowhere; there was Tobias Häusler of Röthenbach, who engaged in all manner of "supernatural things," with which he sometimes had luck, usually failed, but always seemed to pull a nice profit; and then there was the shepherd of Ödern, who claimed much with his rod. "In fact, a number of farmers from the villages freely claim, 'I am a dowser, I also understand this.' "[46] These miners did not consider common farmers to understand the true practice. According to Dietrich and Kuxe, such men swindled unsuspecting miners and degraded what was otherwise a legitimate practice. The authors were, by contrast, "poor miners," and their practice was grounded in nature and mining experience: "Our rods dip by nature, and the techniques come from mining practice" (*Unser Ruthen schlagen aus der Natur, und die Wissenschafften von einem bergmännischen Exercitio haben*). The *Wissenschaft* of dowsing was a function of mining experience. Honorable and legitimate miners and dowsers such as Dietrich and Kuxe "could not believe" what they witnessed at the mines. They pleaded with the officials to protect them and their work from imposters. The Office considered the complaint justified: nine days later, Henckel, Ponikau, and others ordered local shaft managers to favor the formally employed dowsers over their unauthorized competitors.[47]

Mention of Dietrich—now a foreman—appears again in March of 1743, when the *Berghauptmann* at Freiberg (under the *Oberberghauptmann* in the hierarchy), Hans Karl von Kirchbach, shared with

---

[45] "Wenn es noch Berg Leuthe wären, wäre es zu dulden, allein es mischen sich soviel Bergwergsohnerfahrene ein…den es sind mehrentheils Handwergs Leuthe, als der Maurer Knur, von welchem man allerorthen viel muß hören, daß er prave [brave] Leuthe aufsetzet und ein ehrl. Mann dadurch zu Schanden wird gemacht." Wappler, "Wünschelruten-Geschichten," 63.

[46] "Sogar etl. Bauern auf denen Dörffern sagen frey, ich bin auch ein Ruthen Gänger, ich verstehe dieses auch." Wappler, "Wünschelruten-Geschichten," 64.

[47] "…die hierzu bestellten und angenommenen beyden Ruthengeher, ermelten Dietrich und Kuxken, soviel als sich will thun laßen, vor andere gebrauchen möchten." Wappler, "Wünschelruten-Geschichten," 65.

Dresden his skeptical opinion of the dowsing rod, but favorable estimation of Dietrich. The rod clearly had ambiguous status, though Dietrich himself appeared as a diligent servant: "He is at your royal majesty's service as dowser, and has indeed with his expertise performed a number of trials. That dowsers might err, however, is not to be denied; a common outcome and evidence of which is that the rod occasionally deceives. Still, I can humbly affirm that said Dietrich is the best, and for prompt service there has been no better."[48] The Privy Council at Dresden had received a request by the Weimar Office for a "skilled and experienced" dowser, because "such people...are located in the Erzgebirge." Weimar proposed to cover all expenses for such a "sworn and experienced" person for one month.[49] Having received the words from Kirchbach above, Dresden resolved to dispatch Freiberg's favorite, Dietrich, with a respectable stipend of one taler (40 groschen) daily, beginning June 10. This was considerably more than most workers and foremen in the mine, if only a temporary contract. A major official at the liquation works in Grünthal earned 10 taler weekly in 1785 (a *Berghauptman* earned some 29 taler).[50]

The Dowser from Röthenbach

Tobias Häusler, the man accused of magic by Dietrich and Kuxe above, was the same Freiberg dowser we met in the Oberwiesenthal mine reports for the 1730s. We now return to these documents in search of any indication that Häusler's claims would become problematic, since we know that the Mining Office would declare him a fraud. By doing so, it exposed its assumptions and beliefs concerning proper practice, thereby helping us develop our profile of the legitimate dowser.[51]

---

[48] "Er stehet in Ihre Königl. Majt. unsers allergnädigsten Herrns Diensten und Pflichten als Ruthengänger, und hat zwar von seiner Geschicklichkeit verschiedene Proben abgeleget; daß aber auch die Ruthengänger nicht fehlen solten, ist nicht zu läugnen, als welches ein gemeinen Schicksaal derselben, und ein Beweiß ist, daß die Ruthe bißweilen Fallire. Doch kan [ich] die unterthl. Versicherung geben daß gedachten Steiger Dietrich hier der Beste, und in der Eil nicht beßer zu haben gewesen ist." DHSA, Loc. 36302, Fol. 2–2b.

[49] DHSA, Loc. 36302, Fol. 1.

[50] Baumgärtel, *Bergbau und Absolutismus*, 45.

[51] Nummedal has adopted this methodological approach in the case of alchemy. Tara Nummedal, *Alchemy and Authority in the Holy Roman Empire* (Chicago: The University of Chicago Press, 2007).

Nothing unusual appears at first glance, as Häusler proved himself to be a true *Bergverständiger*. He demonstrated a close familiarity with mining, mine surveying, and the earths then 'breaking' at various digs in the region. He referred to veins according to their position (or hour) vis-à-vis the mining compass, and he described earth and rock in the tunnels as either 'hanging' (*hängende*) or 'laying' (*liegende*), as an experienced miner would. Most of his pronouncements were typical: at St. Dorothea, "near the Hirschner dig, a vein will merge from above after 4 Lachter [8 meters], and rich earth will break for 20 Lachter [40 meters]."[52] The only actual mining map (*Riss*) included in this file derived from Häusler's pronouncements for Deep Maria mine, which suggests that he worked personally with the mine surveyor and other officials.[53]

A March 15, 1738 report included a short, laudatory review of Häusler's practice. The author noted that he lived near Freiberg, not 10 miles (*Meilen*) from God's New Fortune. He enjoyed great fame and reputation, "because of his remarkable experience, acquired from an almost forty-year long practice, and especially because of his assays [*Proben*] and duly confirmed claims at various mines, some local and some distant, both in the mining capital of Freiberg and surrounding area, as well as here in the High Mountains [*Obererzgebirge*]."[54] The author referred in this passage to Häusler's general mining experience before zeroing-in on his dowsing, as dowsing for mineral earth was tied squarely to fundamental mining knowledge. The statement also implied that widespread and long-standing employment in Saxony and Bohemia, but especially at Freiberg, was Häusler's mark of distinction. Furthermore, the author suggested that Häusler confirmed his speculations by performing assays himself. The dowser had an amateurish knowledge of assaying, which later correspondence will confirm. This may have added greatly to his authority and rise to fame.

Another paper for March 21 repeated the same lavish introduction for the dowser, adding that his many investigations over a two-week period had been "meticulous and diligent."[55] Again on April 26, we have the very same introduction on Häusler—"40-year long,

---

52 BAF, Loc. 40014, Nr. 74, Fol. 143.
53 BAF, Loc. 40014, Nr. 74, Fol. 108.
54 BAF, Loc. 40014, Nr. 74, Fol. 140.
55 BAF, Loc. 40014, Nr. 74, Fol. 152b.

continuing practice"—and yet again on May 3. Like Schreiber before him, Häusler had become something of a stock figure, and his name appears in at least 15 reports between March and June of 1738 at the Oberwiesenthal jurisdiction alone, amounting to scores of individual pronouncements.

But upon closer inspection we begin to notice certain unusual statements. A selection for March 4 noted that Häusler had predicted that the mines would produce so much wealth, and the town become so well-known, that even its "children's children's children" would live well. The author identified himself here as the mining secretary (*Bergschreiber*), and he noted that he heard this from Häusler's mouth itself.[56] Perhaps the dowser's statement was merely a trope in mining, an example of miner language (*Bergsprache*). But other, more serious exaggerations emerge. The list of pronouncements, from which the mining chart derived, ended with the following claim: that mineral earth containing as much as 40 to 80 Mark of silver (1 Mark= 16 Lot) would be discovered at Deep Maria mine. March 15, the first paper to formally introduce the dowser, also included a claim for God's New Fortune, near Fichtelsberg, for some 16 new veins yielding 14 to 30 Mark of silver, at least one of which would extend some 250 Lachter (500 meters). These exaggerations, while they did not dominate Häusler's accounts, finally warranted a critical remark by a shaft manager on May 30.

The report of May 30, 1738, signed by Gottfried Seltmann, shift manager and account manager at Unterwiesenthal, is unique in the collection. It begins with a title that draws particular attention: "True and Faithful Account." The document is also unique for ending with a witness's testimony. A mining secretary affirmed that the preceding account about Häusler was true and had been accurately related. Seltmann's paper will help explain exactly why this dowser was losing favor before we turn to the formal inquiry into his practice.

It begins with the standard introduction on the "famous dowser from Freiberg," referring also to his "good knowledge" (*Wissenschaft*). Häusler had worked at Freiberg, Marienberg, Annaberg, Schlettau, Bernstein, and now, for Wiesenthal. As the report moves to Häusler's exact pronouncements for Wiesenthal, however, the content and tone

[56] "Welches also aus seines des Ruthengehers Munde treulich wiedergeschrieben worden von Christian Andr. Allmann, Bergschrbr." BAF, Loc. 40014, Nr. 74, Fol. 147b.

changes, as Seltmann relates a curious dialogue he had with the highly recommended dowser. Häusler had claimed that if the St. Stephen tunnel were to continue just 2 to 3.5 Lachter (4 to 7 meters), mineral earth would be found that was, "not very rich." The shift manager, obviously taken aback by such a strange comment, asked how much silver it would contain. A short back-and-forth then followed:

> He said, 'some 2 to 3 Mark.' To that I answered, 'were God to grant that, it would be rich enough, I would be greatly pleased.' To that, he said, 'that is nothing.' If I were to go 10 Lachter [20 meters] further, I would reach a rich main vein that continued for 80 Lachter [160 meters] forever into the depths, and from which ore would break containing 80 to 100 Mark of silver. To such a claim, I shook my head, because it seemed unbelievable to me. Then he asked why I was shaking my head, whether there was something I did not believe. To that I answered: 'if God were to grant but half this wealth I would be quite happy, because such rich ore has never broke at Wiesenthal.' He then said, 'sure, such a noble silver vein has never been exposed [*entblößet*] nor dug out [*gemacht worden*] at Wiesenthal.' But he could assure me, that if such a vein were sunk, the [silver] content would reach well up to 120 Marks. To that I said, 'this would surely be a princely treasure.' He responded: 'whoever owned a share [Kux] when this noble vein were reached would be able to keep himself quite rich with no work.'

Häusler began facetiously by predicting a modest find. As Seltmann pressed the dowser on the quality of the earth, however, Häusler's claims became more extraordinary, as if he wished to end the interrogation by awing the interrogator. But the final comment in the dialogue about purchasing shares raised the central problem about dowsers, and it was far more mundane that we might expect. Häusler claimed that anybody who owned shares (*Kuxe*) here would become wealthy.[57] The problem with a deceptive and self-serving dowser was not that dowsing per se was false, but rather that many people, among them poorer workers and artisans, became shareholders on such speculations. An imprudent dowser might contribute to real losses and poverty. This was precisely the issue to which Seltmann then turned.

[57] The cost of mining was traditionally divided into 128 parts or shares (*Kuxe*), divided among members of the *Gewerkschaft*, or shareholders. Not only merchants and nobles took part, but also common miners and artisans, oftentimes through their collectives (the guild or miners' brotherhood), as well as the local church and other institutions. A Kux normally implied Zubuß payment as well, or support for upkeep of mining irrespective of relative performance.

The issue of shares was widespread in these reports. Among the papers for Oberwiesenthal are scattered concerns that the Central Office in Freiberg had recently levied a new tax on all shares. According to Seltmann, it currently amounted to 5 taler per share, in addition to the 8 or 9-groschen payments required for upkeep (*Zubuß*). This meant that miners and artisans were finding it more difficult to join the shareholding collectives, and the consequences of a poorly performing mine had become more serious. Seltmann still had some 30 shares to distribute that he would, "gladly bring to mining-loving gentlemen and God-loving friends, and complete the shareholding group, so that [he] could back the construction and direct it more strongly." But Seltmann had decided to sell shares with the going tax (amounting to 1 taler, 12 groschen per Kux) rather than the new one, "so that poor people could also purchase, and the shareholding group move ahead, and become complete," and moreover, he did not intend to distribute further shares, "before the dowser's first assays."[58] Judging from his language and decision, it appears that Seltmann wished to protect small-time investors from Häusler. The author ended the 'Relation' by calling attention to his distinguished witness's signature, "so that nobody would doubt the above relation about the dowser, and consider it untrue."[59] That Seltmann took this extra measure is in itself an indication of the reputation that Häusler had been enjoying as a trusted dowser.

We lose track of Häusler for the summer of 1738, but he appears in correspondence between Dresden and Freiberg even before Christoph Dietrich's damning statement of November that Häusler dabbled in magic. On October 11, the barge manager (*Flossmeister*) to Kösen, Georg Harttig, wrote directly to Dresden requesting a reliable dowser. He began by affirming his confidence in the practice. While he conceded that many question its legitimacy, and that "among 100 dowsers, hardly one is found who understands this properly, and knows how to use the rod," Harttig reported nonetheless that a major salt spring at Kösen had been discovered by these means.[60] Few believed that such wealth resided in so desolate and wild a region, and at so great a depth as eighty Lachter (160 meters). Only the rod had proven effective when

[58] BAF, Loc. 40014, Nr. 74, Fol. 141b–142.

[59] BAF, Loc. 40014, Nr. 74, Fol. 142.

[60] "...unter 100. Ruthen Gängern kaum einer gefunden wird, der solches recht verstehet, und die Ruthe zu gebrauchen weiß." Wappler, "Wünschelruten-Geschichten," 74. Bad Kösen is a spa town today.

other treasure hunting methods had failed: "where not the slightest track or other sign or mark was present."[61] After a period of high productivity, conditions at these salt works had then declined, due chiefly to the scarcity of wood. Now Harttig had reason to believe that coal lay in the area, an alternative energy source he lamented having underestimated some twenty years prior. He was asking Dresden to send a certain well-known dowser to search for coal, and preferably before the winter snows complicated matters further. Von Hennicke promptly forwarded the issue to Freiberg, requesting that it dispatch the "skilled dowser" in question, one Tobias Häusler, to Kösen.

The first indication that Freiberg was tightening its grip on the dowser from Röthenbach comes in May of 1739. Dresden asked Freiberg to comment on Häusler's skill, suggest a weekly salary, and formally contract him. Meanwhile, Häusler was to remain in the land and undertake no new service.[62] The Freiberg Office, including Henckel again, the mine master, Hans Karl von Kirchbach, the chief administrator of mines, Johann Friedrich von Ponikau, and the privy councilor, Johann Georg von Wichmannshausen, reported that the dowser's claims were mostly *in*accurate, although on many occasions Häusler had proposed successful digs. They added that Häusler's "Wissenschaft" was convincing and that he was "not inexperienced in Chymicis," or had an amateurish knowledge of metallurgical chemistry and assaying.[63] The former term referred to Häusler's dowsing method or practice, which the authors distinguished here from metallurgy and chemistry. The Office accordingly supported Häusler's continued salary at one gulden (20 groschen) weekly, a common digger's salary, and drafted an 8-point contract, which Dresden subsequently authorized.

The "Provisional Sketch of Instructions for Dowser Tobias Häusler" (*Ohnmaßgebliches Projekt zu einer Instruction, Vor den Ruthen Gänger Tobias Häußlern*) is a code of conduct that articulates the Mining Office's conception of the ideal dowser. After a general statement on Häusler's subordination to the Elector, items 2, 3, and 4 concern proper

[61] "...wo nicht die allergeringste Spuhr oder andere Anzeige und Merckmahle vorhanden gewesen." Wappler, "Wünschelruten-Geschichten," 74.

[62] DHSA, Loc. 36263, Fol. 1.

[63] "...wie Häußlers Angaben überall nicht so accurate eingetroffen, iedoch aber an unterschiedenen Orten erfolget sey...folglich dieses Mannes Wissenschaft nicht zu verachten er aber überdieses im Chymicis nicht unerfahren wäre." DHSA, Loc. 36263, Fol. 3.

use of the rod and the moral conduct of the practitioner. Item 2 is especially revealing for distinguishing between superstitious and legitimate practice:

> Because God the Almighty is to thank for the blessed gift that the dowsing rod dips for him and is used toward the exposure of crevices, mineral veins, feeder veins, and other such things that are hidden to human eyes under the earth; so should he in no way misuse it to the harm of his neighbor, and much less perform superstitious things with it, but rather operate it as a natural means of exposing crevices, mineral veins, and feeder veins.[64]

The Office, like practitioners Dietrich and Kuxe and Hans Wolff above, humbly credited their Maker for granting the ability to dowse. The true dowser did not abuse his powers by applying the rod toward ends beyond his God-given capacity to expose hidden veins. He did not deceive his fellows with strange incantations and prophecies, but rather trusted in a natural correspondence between rod and underground materials. The next item continued with the moral conduct of the dowser, specifying that when Häusler offered his services to others at the mines, he should call on God's help and behave "soberly and moderately," rather than make extraordinary claims. A temperate character (and the authors may have been referring to alcohol), the Office explained, would ensure that Häusler gave proper attention to actual indications of the rod, and guard against any tendency to compel the rod beyond its natural function.[65] Item 4 stressed, once again, that Häusler should refrain from all superstition, avoid making exaggerated claims or "bragging" (*Großsprecherey*), and never deceive his fellows.[66]

The remaining items concerned more practical matters, and illustrate how committed Freiberg had been to this experienced dowser.

---

[64] "Da er Gott dem Allmachtigen die Gnaden Gabe daß ihm die Wundschel Ruthe schlagt, zu dancken hat, selbige aber zu Ausgehung Klüffte, Gänge, Thrümer, und dergleichen, so denen menschlichen Augen unter der Erden verborgen sind, hauptsächlich gebraucht wird: so soll er solche in Keiner Wege zum Nachtheil des Nechsten mißbrauchen vielweniger damit abergläubische Dinge vornehmen, sondern sich derselben, als eines natürlichen Mittels zum Ausgehen der Klüffte, Gänge, und Thrümer bedienen." DHSA, Loc. 36263, Fol. 5b.

[65] "...soll er sich, unter Anruffung Göttlichen Beystandes, und Erhaltung der ihme verliehen Gnade dergestalt nuchtern und mässig verhalten, damit das Nachdencken und nöthige Aufmercksamkeit beym Ruthengehen durch den Trunck nicht verhindert werde, vielmehr sich auf das Ausgehen derer Klufft und Gänge, auch Thrümer, so viel die Wündschel Ruthe anzeiget, gewissenhafft befleissigen." DHSA, Loc. 36263, Fol. 6.

[66] DHSA, Loc. 36263, Fol. 6.

Item 5, which recognized that Häusler had "especial experience" in assaying gold, asked that he carry out any request in this capacity without deception (*in keine Wege...etwas zu verheimlichen*). The final three items (6, 7, and 8) specified Häusler's salary of one gulden weekly, provided for additional funds (12 groschen daily) when travel was expected, and required that he not venture out of the land without formal permission, and live in the Freiberg jurisdiction, to be "at hand."[67]

Problems arose as early as September 1739. The administration learned that Häusler was no longer satisfied with items 6, 7, and 8 of his contract. It reported to Dresden that Häusler wished to receive adequate compensation above his regular salary for his assaying work, which had evidently become an important responsibility; that he found his travel expenses to far exceed his means (Item 7), and suggested that free pass (*Freye Fortschaffen*), for example through tolls, would resolve the matter; and finally, that Häusler was still unable to meet Item 8 and settle in the Freiberg jurisdiction, as he needed to fulfill a rental contract in Lößniz, on Schönberg territory. Aside from these three concerns, Häusler also wanted more freedom to dig without hindrance, for example, from landowners or other jurisdictional offices (*Bergämter*). On this issue, Freiberg responded directly by assuring Häusler that all Saxon jurisdictions would receive formal notice of his commission. Otherwise, Dresden promised Häusler recompense for his metalwork, granted him a continued stay in Lößniz, and offered either additional funds or some sort of letter providing for easier transit. The dowser clearly remained in good standing at Freiberg and Dresden.[68]

A record of his activities over the next few years is missing, but a report of April 1743 to Dresden confirms Häusler's employment by Freiberg and numerous other mining towns, as well as Dietrich's claim that his competitor engaged in "supernatural things." Signed by Kirchbach, Ponikau, Henckel, and the soon-to-be *Oberberghauptmann* and joint founder of the Freiberg Mining Academy, Friedrich Wilhelm von Oppel, among others, the 1743 report on Häusler is a significant account of how Freiberg luminaries distinguished proper and improper use of the dowsing rod.

The report proposed to answer three questions concerning Häusler: what had he produced thus far? What sort of service did he offer? And,

[67] DHSA, Loc. 36263, Fol. 7b.
[68] DHSA, Loc. 36263, Fol. 13–17b.

what could be expected of this service in the future?[69] First, we learn that Häusler had never moved to the Freiberg region, as originally requested, but rather remained at Lößniz, where he resided since 1738. Freiberg reported further that the mining jurisdictions all agreed that Häusler made exorbitant claims, and that his promises had not proven true. Häusler himself apparently admitted to this, but he remained confident in his rod.[70] He also boasted about earlier work at Freiberg, including helping to bring Methuselah mine to surplus, and opening an adit at Three Tigers mine that was instrumental in ventilation and water control. Freiberg addressed these claims, conceding that Häusler did once propose the site for an adit, but it denied that favorable conditions at Methuselah were his doing. It attributed this rather to God's blessing, proper construction, and good management.

In general, the Office reported that the various jurisdictional offices could report nothing reliable in Häusler's practice. At Scheibenberg, for example, "not even the slightest of his claims, which are often completely astounding, at any mine where he has been employed, has proven true."[71] Häusler succeeded instead in ruining investors. At Lößnitz, Häusler once claimed at God's Blessing that a rich vein would be discovered after seven Lachter (14 meters). When the party had already covered twenty-one Lachter (42 meters), Häusler contended that the vein must have been especially deep. Officials determined from this and other cases that, "his rod did not dip at all to crevasses and mineral veins, or, in any case, most indeterminately."[72]

The Freiberg officials became convinced that Häusler was an imposter based on further observations. For example, he did not have the straightforward demeanor of an honest *Ruthengänger*, but showed rather "subtle ambiguity and craftiness" in his dealings. They grew suspicious when Häusler refused to test his ability by the common

---

[69] DHSA, Loc. 36263, Fol. 18.

[70] "...daß Häusler zwar zu reichen Berg Seegen Hoffnung gemacht, zur Zeit aber sich noch weniger effect gezeiget; dahingegen Häusler selbst vorgestellet, daß ob zwar dasjenige was er mit seiner Ruthe...angegeben, zur Zeit noch nicht an allen Orten eingetroffen und der Berg Seegen sich nicht gezeiget, solcher doch noch künfftig zur Erfüllung kommen würde." DHSA, Loc. 36263, Fol. 18b–19.

[71] "Das Bergamt Scheibenberg weiß...anzuführen, daß auf keiner Zeche, wo er adhibiret worden, das geringste von seinem Angeben, welches öffters ganz erstaunlich gewesen, zugetroffen." DHSA, Loc. 36263, Fol. 20.

[72] "...daß ihm seine Ruthe gar nicht oder doch höchst ungewiß auf Klüfft und Gänge schlage." DHSA, Loc. 36263, Fol. 20b.

"experiment" of locating deliberately hidden pieces of metal. The true dowser performed this task "with certainty."[73] The officials also expected Häusler to be familiar with the exact direction or strike (*Streich*) of the veins he discussed: "He drew a good 100 veins on the table and paper according to his supposed dowsing rod, but he could not declare their strike."[74] Häusler, finally, tinkered with magic. A high mining official once tested the would-be dowser at Elterlein by presenting him with a certain mineral stone and asking him where it 'broke,' or where its source lay. Dowsing with mineral fragments in hand was a common practice: like magnetite (lodestone) and iron, the fragment imparted to the rod the virtue of the soil or stone from which it came. Häusler made a prediction of four Lachter (8 meters) from a given spot. He then drew the planets on a table, produced a small lead stone (*ein Stüffchen weißes Bley Erzt*), and actually proceeded, before the assistant to the mine master, to magically see the mines on its shiny surface. "His uncertainty became even more obvious when neither shed, pit, hill, nor tunnel came to sight."[75] This sarcasm expressed the official's skepticism of the magic, although he allowed Häusler to proceed.

Other offices would agree that Häusler's practice was highly problematic. For four years at Annaberg, despite having made great promises and describing in some detail the location of mineral veins, Häusler had apparently not produced one reliable claim. Many were so unbelievable that, "knowledgeable miners began to have suspicions." Since the digging proceeded in vain and at great expense according to his claims, Annaberg concluded that nothing "fruitful" could be expected of Häusler.[76] Geyer, Marienberg, Johanngeorgenstadt, and

---

[73] "...angesehen er nicht in Stande sey, das sonst bey Ruthengängern gewöhnliche experiment, da man das Metall unter die Hüte oder sonst zu verdecken pflegt, mit Gewißheit auszuführen, sondern bey seiner ganzen vermeinten Kunst eine subtile Zweydeutigkeit und Klugheit bezeigte." DHSA, Loc. 36263, Fol. 20b–21.

[74] "Er mahlte zwar nach seiner vermeinten Frag-Ruthe wohl 100. Gänge auf den Tisch und das Pappier, könnte aber ihr Streichen nicht angeben." DHSA, Loc. 36263, Fol. 21.

[75] "Seine Unwissenheit aber würde sich noch mehr offenbahren, wenn er weder Kauhe, Schurff, Halde, und Stolln zu sehen bekäme." DHSA, Loc. 36263, Fol. 21b. A 'Kauhe' (Kuh) was a protective structure erected over the opening of a mineshaft.

[76] "...alleine dieses Angeben ware so groß gewesen daß verständige Bergleuthe iederzeit ein Mißtrauen darrin gesezet, und da der Bau nach seinen Angeben verführet und viel Unkosten angewendet worden...dahero das Berg Amt von Häusllern nichts fruchtbarliches angeben könnte...oder noch von ihm zu erwarten stünde." DHSA, Loc. 36263, Fol. 22.

Ehrenfriedersdorf all reported similar experiences with the ubiquitous Häusler.

The Freiberg Office ended the report by charging Häusler with, "gross ignorance, unabashed boasting, and spreading nonsense."[77] It answered the three original questions: Häusler's track record as a dowser was poor, his practice was suspect, and his prospects were questionable. Dresden accordingly revoked his salary. To continue patronizing Häusler would bring only "malicious gossip and real harm" to Saxon mining.[78]

Profile of the Hired Dowser

The case of Tobias Häusler, including the 'Provisional Sketch of Instructions,' is our strongest indication that the mines were a testing ground for dowsing in the eighteenth century. Confronting the illegitimate dowser required articulating proper practice. Freiberg was famous for its dowsers, and the would-be practitioner who sought legitimization and status for his skills appealed to the center of mining operations in the Erzgebirge. We can now use the preceding accounts, and the Spitzberg salt expedition to generate a profile of the hired dowser. What sort of person was he? What sort of social status did he possess? What distinguished the true dowser from the imposter?

We might begin with the obvious fact that this person was male. The hired dowser was a hearty miner. Like men who dabbled in natural magic, dowsers assumed control over and even directed the forces of nature. Women were rarely granted the ability to control natural forces, as was most evident in the witch-hunts, when male authorities feared such power and identified it with diabolism. The official dowser demonstrated a familiarity with mineral earths that was typical among miners but uncommon among women, who had greater experience with surface-level vegetation. He should not have a weak constitution that would allow him to be overrun by natural forces, as physicians described the possessed or mad,[79] but demonstrate the

---

[77] "Grobe ingnoranz, unverschämte Auffschneiderey und Windmacherey." DHSA, Loc. 36263, Fol. 22b.

[78] "Durch demselben dem Bergbau übele Nachrede und würckliches Nachtheil zuwachsen werde." DHSA, Loc. 36263, Fol. 24.

[79] Humanist physician and legal thinker Johann Weyer argued that accused witches, 'melancholic' and overcome by black bile, were physically and mentally incapable of

mental concentration and know-how to lead important officials to nature's hidden treasures. At the mines, the dowser might shake, but not uncontrollably, as the sickly Hans Wolff had done. When Göbels became sick in 1714, Voigtel and Schneider noted that sickness might compromise accurate dowsing. According to the historian of magic and witchcraft, Eva Labouvie, male magic tended to involve more aggressive control over mysterious forces than female magic, and it often concerned material-oriented activities such as treasure-hunting and dowsing.[80]

The sphere of mining was male. Women had performed a number of functions in mining into the 1580s, but this changed with demographic expansion and inflation, and a transition to the deeper-level, capital-intensive mining that we have been discussing. The mining brotherhoods (*Knappschaften*) began to dominate the organization of labor, and they excluded women. Mining, as other trades, cultivated a highly masculine ethic after the sixteenth century. Women may have wielded the rod for smaller-scale treasure hunting in the village, much as they functioned as popular healers, but female dowsers were the exception that proves the rule. Martine de Bertereau, the wife of an important baron and mining consultant to Louis XIII of France, had been dowsing since her teens when her husband and his work force undertook a mineral survey of the kingdom in 1626. In a book addressed to none other than Cardinal Richelieu, the Baroness claimed to have discovered 150 mines by her dowsing rods.[81] In England, Lord Byron's mother-in-law, the noble Lady Milbanke, was an avid dowser and correspondent with the editor of the Royal Society, Charles Hutton. After he translated and published a work that rejected dowsing, Lady Milbanke challenged Hutton to test her practice. He was satisfied that the motions of the rod were real, but unable

---

making a legal contract or pact with Satan. Later judges and physicians returned to this argument, which was one factor behind the decline of persecutions. Erik Midelfort, *A History of Madness in Sixteenth-Century Germany* (Stanford: Stanford University Press, 1999), 206–11.

[80] Eva Labouvie, "Perspektivenwechsel: Magische Domänen von Frauen und Männern in Volksmagie und Hexerei aus der Sicht der Geschlechtergeschichte," in *Geschlecht, Magie und Hexenverfolgung* (Bielefeld, 2002), 46–7. See also her, "Men in witchcraft trials: towards a social anthropology of 'male' understandings of magic and witchcraft," in *Gender in Early Modern German History*, ed. Ulinka Rublack (Cambridge: Cambridge University Press, 2002).

[81] Martine de Bertereau, *La Restitution de Pluton*, 1640. Both husband and wife were incarcerated on suspicion of witchcraft, and died in prison.

to offer an explanation.[82] But neither elite women nor those working as popular soothsayers and treasure-hunters worked for mining jurisdictions.

As for his personality and comportment, the hired dowser walked a fine line between demonstrating divine favor and conducting himself with evident humility and deference before officials. One legitimate dowser was the "old and honest" carpenter (like Jesus himself) that Beyer mentioned, perhaps recognizing his predecessors in the Bible, recalling the staff of Moses, or the famous sermons of Johann Mathesius, where Adam mined "with or without the rod." Granted the privilege to dowse, the hired dowser accordingly made modest and reasonable claims, unlike those of Häusler; and he did not presume familiarity with learned authorities on the subject, as did Hans Wolff. The official dowser was sure of his abilities and confident, but not arrogant. He grounded his *Wissenschaft* on divine favor and experience at the mines, as the Häusler contract stressed he should, rather than on bookish learning or distinctive powers. Dietrich and Kuxe upheld their "miner experience" over their inexperienced competitors. Called into the Dübener Moor to confirm salt springs, Meinel was said to possess great "miner wisdom," but there was no indication that the peasant dowser, Hans Fischer, had such experience. The mining reports all emphasized the many years of experience that men like Knorr, Schreiber, Göbels "of Freiberg," and Häusler had in mining.

The dowser was reliable, at-hand, and forthcoming with his knowledge and abilities, even available for examination like Dietrich. Most of the dowsers mentioned in the mining reports were locals formally employed. They should be transparent, sedentary, and submissive to authority, in other words, rather than crafty, elusive, or on the move, like Junghans or Häusler. One of the most common tests, which both Wolff and Häusler rejected, involved hiding coins and requesting the would-be dowser to locate, if not identify each piece.

The hired dowser was not a mine surveyor, but he was no digger or peasant either, like Hans Fischer. He was a foreman like Caspar Fischer or Christian Dietrich, or even a mine manager like Johann Meinel. Dietrich and Kuxe complained that common farmers purported to

---

[82] In Christopher Bird, *The dowsing hand: the art of searching for water, oil, minerals, and other natural resources or anything lost missing, or badly needed* (New York: Dutton, 1979), 111.

dowse, and the true dowsers asked the Freiberg Office to formally distinguish them from the riffraff. Schreiber and Voigtel in the Dübener Moor quickly distinguished between Hans Fischer and Göbels. In a number of cases, dowsers from Freiberg also exposed the falsity of commoners' claims, underscoring a fundamental difference between popular treasure hunting and official dowsing. The dowser's wage reflected his distinct social position; it varied, but tended to approximate that of a foreman. On special commissions, the dowser earned even more—Dietrich earned 1 taler daily in 1743 for a particular job. Often already formally employed at the mines, the dowser understood the layout of veins according to the mine compass and spoke the language of mining: "morning vein" (*Morgengang*), "evening vein" (*Spätgang*), "strike" (*Streich*) and "breaking" (*brechen*) of earth. He presented himself as a *Bergverständiger*, or honorable and experienced miner. Contracted for a single commission or given a yearly salary, and often under oath, the dowser was highly visible and integrated into the mining community, unlike the alchemists who worked more secretly both at town and court.[83] Abraham Schönberg requested that the dowser be paid like others at the mine in 1709, unlike the alchemists, astrologers, healers, and magical treasure hunters, both male and female, who also lived on the mountain, but who offered their services in less visible and formal spaces.

This is not to suggest that the formalization of dowsing in Saxony was free from controversy or that the legislation ever addressed dowsing as it did assaying and mine surveying. It was precarious business to propose a dig, which, while it might confer great visibility to the dowser, also assured a struggle for confirmation. Like a courtier, prospects were high for the promising dowser, such as Dietrich, who was dispatched to Erfurt at high salary and joined the Elector to Poland. But the dangers of falling from grace were always present, as in the case of Häusler, although this was a far cry from the counterfeiting charge that befell the alchemist in the sixteenth century. However often officials dropped the derogatory '*Wünschel*' (wishing or dowsing) in their

[83] Like many other European princes, the Saxon electors were deeply interested in alchemy and oversaw the practice like a trade secret, keeping the work from public view. Court festivals that showcased the Saxon mines did, however, feature alchemical and astrological imagery. "Many of the alchemical writings were kept behind closed doors in the laboratories in Dresden… But the public face of alchemy was its connection to mining." Watanabe-O'Kelly, *Court Culture*, 107.

correspondence over the rod, and however formal the handshakes and contracts, this remained a highly contested practice among scholars and officials that would never assume the level of integration into mining that warranted formal attention in the mining legislation. Dowsing moved to the very edge, not the center of official concern in the eighteenth century.

Finally, other than appeals to divine grace, there were only scattered references to the mechanism at work behind dowsing in the salt expedition, and mining reports and official correspondence we have reviewed above. We know from mining books and will see again in Chapter Five that theories of dowsing were readily available—mineral fumes, sympathetic attraction, occult qualities, corpuscularism—but practitioners seemed more intent on defending their moral character and experience at the mines than developing or defending theory. They wished to be the sort of miner one could trust, not the sort of intellect that could explain why dowsing worked. On two occasions, the officials in charge at Spitzberg Hill asked their dowsers to concentrate their minds in a particular way, which would suggest a mental explanation for the dip of the rod; but the order to hold a mineral fragment (*Geschiebe*) to the rod suggests also an affinity or mineral vapor theory. Beyer and Rößler had discussed the practice of holding mineral rocks to the rod in their mining books. Perhaps mental influence and mineral vapors were not mutually exclusive forces. For whatever it is worth, Beyer had said in his surveying manual that critical thought interfered with dowsing practice. Häusler was credited for his "*Chymicis*," but it was not a flawed theory or chemical practice that secured his downfall, but rather his pride, exaggerations, and an inability to demonstrate the proper location of veins. Hans Wolff was ridiculed for possessing books of Valentine (Thölde) and Paracelsus, although we know that both authors had speculated about mineral vapors.

Still, we detect a widespread assumption among our officials and practitioners that the rod worked by natural means. That belief never achieved the sophistication of Thölde's text or the mining books of Beyer, Rößler, or Löwel, but invisible rays or vapors of some sort caused shaking of the body and pulled on the dowsing rod. The stronger the vapors, the stronger the pull. Dietrich and Kuxe carefully distinguished their practice from the "supernatural things" performed by Häusler, such as gazing into a polished stone. But his practice was not entirely supernatural: he too was once asked to dowse with a mineral fragment in hand that would have increased the attractive virtue of his rod to the earth below. The open-minded but skeptical officials observed their prospective dowsers closely and allowed for a large amount of idiosyncrasy—Wolff spit on his rod—and accumulated experience, but they assumed a natural correspondence between rod and earth.

Diviners and Scholars

We have captured a period in the history of mining science at Freiberg when '*Wissenschaft*' retained craft associations and referred to experience in mining, God-given abilities, and knowledge of the hidden properties of nature—to the *divining* of information. Mining officials and other elites looked to the miners and locals assumed to possess the requisite knowledge of the earth. In theory removed from magic and superstition, divining became a visible component of official mining. But the critical analysis of men like Johann Voigt, who examined 'Shaking Hans' in 1713 and authorized the Spitzberg expedition, gave way to more hostile opinions among Mining Academy scholars after 1765. The community of earth science specialists that gathered at the Academy proposed a mining science at some remove from *Bergverständige* like Christoph Dietrich or Johann Meinel, and more in line with an international Republic of Letters. A higher standard of professionalism and education had entered Saxon mining after the economic downturn of the Seven Years War. That standard would challenge the pragmatic approach to dowsing that characterized Nicolaus Voigtel and Henckel's day. Before turning to the Enlightenment in Saxon mining and the new Academy, we should introduce scholarly interest more broadly in miners and miner beliefs. Natural philosophers prior to the Academy did not unanimously cast the dowsing rod to the realm of worthless superstitions, any more so than the theologians had condemned it.

CHAPTER FIVE

THE MURDEROUS MATTER: DOWSING AND NEW SCIENCE

"Just as the effluvia in the lodestone cause it to pull the same material strongly to itself (because the lodestone is the best iron ore)...so do the same effluvia proceed from minerals and metals and unite with the effluvia of the dowsing rod."[1] These were the words of a Brandenburg jurist who explained that mineral effluvia consisted of particles that filled tiny holes along the rod and pulled it mechanically downward. The explanation was grounded in contemporary mechanical theory coming from France, and it developed a long-standing association between the compass and dowsing rod. Corpuscularism was one of many learned theories to emerge during the Scientific Revolution to account for the mysterious motions of the rod.

The major learned debate over dowsing of the seventeenth and the eighteenth centuries was no longer whether the rod worked by diabolical or natural means, but whether it was cutting-edge science (*Wissenschaft*) or popular superstition (*Aberglaube*), a distinction that had social, epistemological, and even gendered meaning to those who used it. Diverse scholars of the Scientific Revolution and early Enlightenment had great interest in the earth and minerals. Inevitably, their works had major impact in European mining, as the mines became an important site for the acquisition of mineralogical and stratigraphic knowledge. Early chemistry found an audience at the Swedish Board of Mines already in the seventeenth century, and aspiring officials at Freiberg studied chemical mineralogy after the 1730s, and geognosy, paleontology, geophysics, and other new sciences by century's end. Great Britain, Austria, Russia, and Italy also developed sites devoted to the union of new science and mining.[2] What was the

---

[1] Johann Philipp Büntigen, *Sylva subterranean* (Magdeburg, 1693), 100.

[2] Ezio Vaccari, "Mining and Knowledge of the Earth in Eighteenth-Century Italy," in *Annals of Science* 57: 2 (2000). Ernst P. Hamm, "Knowledge from the Underground: Leibniz Mines the Enlightenment," *Earth Sciences History* 16: 2 (1997), 77–99. Hjalmar Fors, "Occult Traditions and Enlightened Science: The Swedish Board of Mines as an Intellectual Environment 1680–1760" in Principe, ed., *Chymists and Chymistry*, 239–52. Theodore Porter, "The Promotion of Mining and the Advancement of Science: The Chemical Revolution in Mineralogy," *Annals of Science* 38 (1981), 543–570.

fate of dowsing within the scholarly community? Did savants, natural philosophers, and experimenters develop new *Bergwissenschaft* to replace that which the mining books of Rößler, Beyer, and others had presented, in which surveying and dowsing could appear as complementary practices? Did the mineralogical knowledge of a Descartes, Steno, or Leibniz undermine mineral vapor theory?

This chapter surveys the interest select scholars across Europe had in minerals, mining, and dowsing during the Scientific Revolution. Some of the earliest contributions to the *Philosophical Transactions of the Royal Society* will introduce the topic, as they illustrate how natural philosophers gathered information about miners, what the important issues were, and how German mining experience informed practice abroad. In France, promoters of the new Cartesian philosophy claimed to account for the case of a peasant dowser who tracked down the culprit of a murder in 1692: the man had dispersed corpuscles that the dowser had mechanically detected. So presumptuous was this extension of the mechanical philosophy, said its opponents, that one issued a sharp response. Johann Gottlob Zeidler (1655–1711) was a writer living in Halle, Germany, the center of an anti-Cartesian and vitalist movement spearheaded by Georg Ernst Stahl among others. Zeidler's treatise on dowsing and the 1692 Aymar case, prefaced by the famous philosopher and fellow anti-Cartesian, Christian Thomasius, became an influential statement among those who would prefer a more spiritualist, or mental explanation of the dowsing rod's motion.

Lest we assume that dowsers everywhere found the sort of audience that Aymar did, Leibniz represents another, more critical trend. Unlike many of his contemporary philosophers, Leibniz actually worked in mining to revamp practice and introduce new earth science. His time in the Harz Mountains, while it informed Leibniz of prospecting beliefs, also exhausted his patience for such superstitions as underground spirits and dowsing rods, which the *Philosophical Transactions* had entertained, and which the Cartesians seemed to dignify. In his writings on mining and earth science Leibniz, like later Enlightenment philosophers, distinguished between 'science' and 'superstition.'

Given these different approaches to mining and dowsing, what sort of mining science would the Academy at Freiberg advance? How indebted to the *Bergverständiger* like Beyer would it be? This chapter raises these issues and begins providing answers, but leaves a more focused treatment of the Academy for Chapter Six.

Digging through the Philosophical Transactions

German miners and metallurgists, among other artisans and merchants from the Continent, had been providing their services in England for centuries. German workers were at work in Cornish mines as early as the thirteenth century, and German financiers helped found the Society of Mineral and Battery Works and the Society of Mines Royal, hiring experts from home to spearhead England's copper mines in the 1560s.[3] Queen Elizabeth's court showed great interest in German metallurgical experts, particularly from Saxony.[4] The famous merchant family from Augsburg, the Höchstetter, remained active in Keswick into the seventeenth century, and in 1618 London financiers consulted them concerning new operations at Cornwall.[5] When King Charles II sought to revive the lead and tin mines at Mendip Hills and Cornwall, a continuing interest in German expertise was evident at the newly chartered Royal Society of Science in London.

The Restoration of monarchy after the Civil War marked the beginning of the Royal Society, which received its charter from Charles II in 1662. The new Society showed a marked interest in mines, minerals, and the earth from the start. Boyle was a member of the Company of Mines Royal, which subsidized major mining operations. Already in 1660, he published a piece on the purported growth of metals exposed to the atmosphere.[6] Hooke discussed fossil remains and the structure of the earth as early as 1687 in his 'Lectures and Discourse on Earthquakes.' The first secretary of the Society, Henry Oldenburg, actively sought information about mining and minerals from his native Germany, within his extensive correspondence.[7] Newton's speculations

---

[3] Raingard Esser, "Germans in Early Modern Britain," in *Germans in Britain Since 1500* (London: The Hambledon Press, 1996), 17–27. Eric H. Ash, *Power, Knowledge, and Expertise in Elizabethan England* (Baltimore: The Johns Hopkins University Press, 2004), especially Chapter One. Maxwell Bruce Donald, *Elizabethan Copper: The History of the Company of Mines Royal, 1568–1605* (London: Pergamon Press, 1955).

[4] Deborah E. Harkness, *The Jewel House: Elizabethan London and the Scientific Revolution* (New Haven, CT: Yale University Press, 2007), 169–180.

[5] Esser, "Germans in Early Modern Britain," 26.

[6] "Observations about the Growth of Metals in their Ore Exposed to the Air," in Michael Hunter and Edward B. Davis, eds., *The Works of Robert Boyle* (London, 1999–2000), vol. VIII, 145–52. On Boyle's related interest in alchemy, see Lawrence M. Principe, *The Aspiring Adept: Robert Boyle and his Alchemical Quest* (Princeton: Princeton University Press, 1998).

[7] See "Koch to Oldenburg, 18 November 1664" [Letter #354] and "E. Browne to Oldenburg, 10 April, 1669" [Letter #1145] in A. Rupert Hall and Marie Boas Hall, eds.,

on the geochemical processes of the earth have recently been studied.[8] The "Articles of Inquiries Touching Mines," published in the *Philosophical Transactions of the Royal Society* in 1666, are more formal steps that experimental natural philosophers made into the realm of mining and prospecting beliefs. While the collection represents the Society's ambitious aspiration to possess and control all things mineralogical, it also proves how nascent that wish remained, and shows continuing dependency on the Continent.

Roy Porter once characterized the Articles as a "random accumulation" of queries that "still embod[ied] Renaissance cosmology,"[9] but we consider it a clear statement of formal interest into mining, including the prospecting practices with which we are concerned here. It comprised some one hundred proposals on mining, touching on such issues as the health of workers, the politics of mining, geography, prospecting, and the generation of metals.[10] According to the publisher, Oldenburg, the Society President (The Viscount Brouncker) composed the list in hopes that certain of his foreign guests would carry the inquiries to their native lands, and report back to London. More locally, court chaplain and Society fellow Joseph Glanville recorded his correspondence with a "Person living near the Mendip Mines" and with "some very experienc'd Mine-men" in Number 28 (1667) and Number 39 (1668) of the *Philosophical Transactions.*

The author divided the Articles into six sections, following the model of mining books such as Georg Agricola's *De re metallica*, which mirrored the mining process itself, proceeding from landscape and natural resource issues, to digging and tunneling, to machines and extraction, and finally to assaying and smelting. In the first section, 'The neighbouring Country about the Mines,' Question 6 asked whether "mineral and subterraneous Steams" from underground caused

The Correspondence of Henry Oldenburg (Madison: The University of Wisconsin Press, 1965–86), vol. I-XIII.

[8] William R. Newman, "Geochemical concepts in Isaac Newton's early alchemy," *The Geological Society of America Memoirs* 203 (2009), 41–49.

[9] Porter, *The Making of Geology*, 18.

[10] "Articles of Inquiries Touching Mines," *Philosophical Transactions* 19 (1666), 329–343. "Answers to some of the Inquiries formerly publish'd concerning Mines," *Philosophical Transactions* 2: 28 (1667), 525–527. "Additional Answers to the Queries of Mines," *Philosophical Transactions* 3: 39 (1668), 767–771. On the *Philosophical Transactions* and wider Republic of Letters, see Rhoda Rappaport, *When Geologists were Historians, 1665–1750* (Ithaca, NY: Cornell University Press, 1997), Chapter One.

particular meteorological phenomena such as humidity, fog, or changes in temperature. This is an indication that *Witterungen*, as they were called in Germany, had an English equivalent. Agricola had discussed the matter in Chapter Two of *De re metallica*, and later mining books included lengthy discussions. Glanville confirmed that the Mendip Hills were indeed moist, cold, and foggy, and "'tis probable, it may arise from the Mineral and Subterraneous Steams." He added that snow, frost, and dew "stay upon Mendip longer," rather than evaporated, which would suggest that warm mineral vapors or heats were absent. However, in the second report of 1668, Glanville related that, "Snow and Frost near the Grooves melt quickly, but continue long at further distance," which suggested that mineral heat indeed had an effect.

The third section, 'The Signs of Mines,' demonstrated that digging beliefs were a major concern for Oldenburg and the Society. The author distinguished two kinds of signs: surface-level and subterranean. Surface-level signs were almost identical to those given in the German sources, and included whether mineral fumes discolored or wilted trees and plants, whether waters washed mineral fragments away, whether the waters themselves had a distinctive consistency or taste, whether the dowsing rod dipped to mineral ore, and whether frost and snow melted above mines. Concerning mineral fumes, Glanville corroborated that at Mendip the trees "have their tops burnt, and their leaves and out-sides discoloured, and scorched with the Wind, and grow to no bigness or stature." Articles 17 and 18 echoed Thölde (Basil Valentine) by associating mineral vapors with the dowsing rod: the author asks whether "mists" rising from the earth signified minerals, and "[w]hether the *Virgula Divinatoria* be used to find out the Veins of proposed Mines; and, if it be, with what success?" These were followed by number 19: "What other Signs above ground afford probability of Mines, or Direction for following a Vein over Hills, Valleys, Lakes, Rivers, Etc."

Addressing Articles 17–19, Glanville exposed contradiction or inconsistency in the information he gathered at Mendip. His first informant of 1667 reported, "That the *virgula divinatoria* hath not been known to have been seen used in these parts. That there are no certain signs above ground, that afford any probability of a Mine." By contrast, the second report demonstrated that workers believed both that particular signs indicated mineral ore, and employed the dowsing rod, if cautiously: "Sometimes, when a Mine hath been very near the surface, the Grass hath been yellow and discolour'd. Some have made use of the *Virgula divinatoria*; but the Experienced Work-men have not value for it; yet they say, when the Mine is open, they may guess by it [the rod], how farr the Vein leades."

The Inquiries then turned to underground signs of mineral ore, and asked whether distinctive earth, stones, water, or heat surrounded mineral ore, and whether any other such signs disclosed the exact type, quality, or depth of minerals. Among other bits of popular knowledge, the miners of the second report told Glanville that white, yellow, or compound earths were "leaders to the Country," or sure signs of rich ore, whereas black stone was a bad sign. Brittle clay meant that mineral ore lay in proximity, whereas "the tough is not leading."

The fifth and sixth sections of the Inquiries are also of interest here. In section five, 'The Nature and Circumstances of the Ore,' the author asked whether metals "grow like plants" (number 51), and whether ore developed over time into noble metal, what he called "Maturation of the Mine" (number 61). This section also included an explicit reference to Mathesius (number 65), clearly an authority across the Channel, who addressed the generation of ore and prospecting techniques. Section six, 'The Reduction of the Ore into Metal,' inquired into matters pertaining to assaying and metallurgy, such as number 67, which asked whether mercury was used to extract metal from baser material (amalgamation). The technique had been known on the Continent for at least a century.

A list of eleven 'Promiscuous Inquiries' that did not fit neatly into the prior categorization followed the six sections. It included another reference to Mathesius, an inquiry into the substance "German Naturalists call Ghur" (a sulfur/mercury mixture Mathesius and other German sources believed to be instrumental in the generation of ore), further supposed effects of mineral vapors on local vegetation, and a query about mining spirits. Promiscuous Inquiry number 6 reads: "Whether the Diggers do ever really meet with any subterraneous Daemons; and if they do, in what shape and manner they appear; what they portend; and what they do, etc.?" Joseph Glanville posited these questions to the miners of Mendip in 1668: "Concerning subterranean Daemons, they have never seen any, but sometimes have heard knockings beyond their own Works, which, when follow'd by them, have afforded plenty of Ore." The Cornish called the spirits 'knockers,' or 'tommyknockers.' The folklore for Saxony is also full of tales of strange banging underground: At Naila, near the southernmost tip of the Erzgebirge, some miners only began working when the *Bergmännlein* who lived in the tunnels banged his hammer three times. The mine was called '*Waldhammer*' (Forest Hammer).[11] Georg Agricola gave the most famous description of mining gnomes in *De re metallica*.

Popular digging knowledge was featured again in "An Account of some Mineral Observations touching the Mines of Cornwal[l] and Devon" (1671) of the *Philosophical Transactions*. The publisher did not give the author or informant of the report, referring to him only as "an inquisitive person, that was much conversant in those Mines." Using miner terminology rather than substituting learned vocabulary, "for avoiding of superfluous words and needless circumlocutions," he began with the "Art of Trayning a Load." This was prospecting, or *Schürfen* in Germany, methods for discovering the location of mineral resources.[12]

'Trayning' was based on a few basic assumptions about the earth's history. The Cornish and Devon miners believed that at some primordial time mineral veins and lodes lay near the surface of the earth. During the Creation and again during the Flood waters carried away exposed sections of mineral veins and lodes from surface-level earth, called the "shelf," "fast countrey," or "ground." Both the earth that immediately surrounded the minerals ("grewt") and mineral fragments ("shoad," "*Geschiebe*" in Germany) settled in valleys and riverbeds, sometimes miles from their points of origin. In search of alluvial shoad, the prospectors studied the landscape, the color and nature of earths, and embedded rocks beside streams and rivers. Especially after storms and other disturbances, they inspected newly exposed rocks and earths, determined their weight and porosity, and crushed the material in water ("vauning"). Like Hans Uttman, author of the 1601 *Bericht*, and others in his tradition, the English trainers also believed that the smoother the shoad, the further it must have traveled from its vein of origin. Having thus determined a promising location, the men then panned the waters for additional shoad before making exploratory digs ("essay-hatches") at determinate intervals up the hill.

Aside from these assumptions about Genesis and the shoad, the prospectors also used the dowsing rod. The author expressed his

[11] Heilfurth, *Bergbau und Bergmann*, 399.

[12] "An Account of some Mineral Observations touching the Mines of Cornwal[l] and Devon; wherein is described the Art of Trayning a Load; the Art and Manner of Digging the Ore; and the Way of Dressing and of Blowing Tin: Communicated by an Inquisitive Person, that was much conversant in those Mines," *Philosophical Transactions* 6: 69 (1671), 2096–2113.

learned skepticism, though he clearly had knowledge of practitioners. He claimed that overly zealous dowsers served only to frustrate trainers, whereas "experimented rules" for locating mineral deposits gave more reliable information.

> Although the *Virgula divinatoria* of some few (whose success I am ignorant of) hath been employed for finding the orifice of a proposed Mine... yet because they [the few] are rather nice, than needful, and not sufficient for what they are urged by some, unless it be to cause the over-curious but unskillful Trainer to desist from a farther search after what by such fallible curiosities may seem not to be, but yet by the before-mentioned and daily experimented rules may easily be discovered.[13]

The 'Articles' provide a snapshot of the Royal Society's interest in mining and prospecting beliefs. And yet, the smooth and precise style of the *Philosophical Transactions* glossed over the social divisions that existed between scholars and lower-status craftsmen, which became evident when the former condescended to approach the workshop or mine. Responding to a list of official queries, a trusted member or friend of Boyle's experimental community consulted informants in the field (about whom we learn little) and relayed this material back to the Royal Society, where it might be published as a response to the query. This paralleled the relationship between experimental natural philosophers like Boyle and their technicians in the laboratory, who had more immediate knowledge of the instruments, but who generally remained 'invisible' in the generation of experimental knowledge.[14] A social barrier between savants and diggers was obvious to the professor of mining and chemistry at St. Petersburg, Johann Gottlob Lehmann, who claimed that certain scholars considered it ignoble to "sully themselves, to don such strange raiments, to crawl about with such a wretched and sometimes coarse people—as one occasionally finds among one or another miner." He added that, "going underground...is not for every scholar."[15] As Lehmann might have predicted, Leibniz, however close his interaction with miners and officials in the Harz Mountains, actually disliked entering the mines.[16]

---

[13] "An Account of some Mineral Observations," 2101.

[14] Shapin, *A Social History of Truth*, Chapter Eight.

[15] Quoted in Hamm, "Knowledge from Underground," 86.

[16] "A fact that couldn't have won him much admiration from miners." Hamm, "Knowledge from Underground," 86.

Before studying Leibniz's proposals for mining and dismissal of prospecting traditions, we should turn to France and examine the most powerful learned theory of dowsing of the Scientific Revolution. However short-lived, the corpuscular theory was a suggestive alternative to mineral vapors, and it incited a strong reaction among anti-Cartesians in Germany that would echo through the halls of the Freiberg Academy. It all began with the curious case of a peasant dowser named Jacques Aymar.

The 'Murderous Matter'

In *Principles of Philosophy* (1644), Descartes argued that the universe was composed entirely of matter in mechanical motion. Sulfur was a composite of "extremely tiny and very flexible branching particles," and salts were "sharpened like swords."[17] The major natural philosophers to develop this line of thinking in mineralogy were Christian Huygens, Nicholas Lemery, Jacques Rohault, Nicolaus Steno, Robert Hooke, and Robert Boyle. According to Steno in the 1669 *Prodromus*, translated by the Royal Society in 1671, minerals should be studied as collections of irreducible particles with billiard-ball-like properties. Hooke speculated that "those steams, which seem to issue out of the Earth" were composed of particles of determinate quantity and weight. He imagined that a barometer would determine "what Minerals lye buried under the Earth, without the trouble to dig for them."[18] His master, Boyle, pursued similar arguments in his 1672 *Essay about the Origine and Virtues of Gems*, in which he discussed minerals as part of his larger corpuscular philosophy. A corpuscular theory of dowsing was bound to follow.

In the summer of 1692, a local judge, intendant, and head of police granted a farmer and dowser from the village of Saint-Veran in Dauphiné permission to investigate the July 5 murder of a wine trader and his wife in Lyon. The bodies had been found in the house cellar, but the unidentified murderer was still at large. Jacques Aymar arrived with a dowsing rod and the unusual claim that, if exposed to the cellar, his rod would bring investigators to the culprit. Aymar was shown into

[17] Quoted in Lauden, *From Mineralogy to Geology*, 33. See also Oldroyd, *Thinking about the Earth*.

[18] In Lauden, *From Mineralogy to Geology*, 33.

the cellar, where he immediately suffered paroxysms and shortness of breathe, as the rod dipped strongly to the exact spots where the bodies had lain. Aymar then led the party on a wide-ranging pursuit through the city and countryside, pausing at the house of a gardener and a local bar, crossing the Rhône River, arriving at a military camp, and detecting three distinct trails. The men then apprehended one of three presumed perpetrators at a Beaucaire jail, a nineteen-year old hunchback named Joseph Arnoul, who, aside from petty theft, now confessed to meeting two murderers in Lyon after the fact. Encouraged by their progress, the men continued the manhunt, which ended finally at the Italian border on the shores of Genoa, where the party lost their jurisdiction. Duly identified by witnesses, Arnoul confessed to having kept guard while his two partners committed the crime. Authorities broke him on the wheel on August 30.[19]

News of these events spread quickly through France, filling pages of the popular *Mercure Galant*, and letters among the French elite, who generally accepted that Aymar had solved the crime, but differed widely in their explanations.[20] Some scoured the Bible and classical works for all references to dowsing; others looked to the stars and their effects on the rod, as well as drawing nativities for Aymar; still others saw only diabolical influence in the whole affair. The head of the school of medicine in Lyon supposed that dowsers were like human magnets, and that their rods functioned like needles of a compass attuned to the vapors of other individuals.

French natural philosophers such as Nicolas Malebranche, Pierre Garnier, Pierre Chauvin, and Pierre le Lorrain de Vallemont, who claimed to examine Aymar himself, discussed the fashionable new philosophy: had the culprit exuded from his skin microscopic particles that Aymar somehow drew with his rod or extended arms, like the eye captured corpuscles of light? Would these particles remain in the air despite the atmosphere? Garnier coined the term "murderous matter" to suggest this possibility.[21] The philosopher and priest, Malebranche,

---

[19] Michael R. Lynn, "Divining the Enlightenment, Public Opinion and Popular Science in Old Regime France," *Isis* 92 (2001).

[20] Much of the debate was collected by the priest of the Paris Oratory, and Malebranche correspondent, Pierre Lebrun, in *Lettres qui Découvrent l'Illusion des Philosophes sur la Baguette, et qui Détruisent Leurs Systemès* (Paris, 1693).

[21] Cristopher Bird, *The Divining Hand: The Art of Searching for Water, Oil, Minerals, and other Natural Resources or anything Lost, Missing, or Badly Needed* (New York: Dutton, 1979), 102. See also Barrett and Besterman.

rejected the theory on the grounds that material emanations decreased with distance, whereas the rod seemed to function better for objects underground rather than those in immediate proximity.[22] In fact, Malebranche used the Wittenberg scholar, Sperling, to argue that the entire affair was a species of witchcraft. The first major publication on the subject of Aymar was *La Physique Occulte, ou traité de la Baguette divinatoire* (1693) by Pierre de Le Lorrain, Abbot of Vallemont, which the Cardinal of Grenoble placed on the Index of Forbidden Books in 1701. It had been translated into German in 1694, however, eliciting a number of German responses. Leibniz, for one, was quick to decide against Aymar.[23]

We can use Vallemont to study the Cartesian theory, before turning to the response from Halle. The abbot claimed to have examined Aymar's abilities twice daily for one month in Paris, remaining convinced that the rod turned in his hands before criminals. Vallemont began his account by referencing the eighteenth of the Society's 'Articles of Inquiries touching Mines,' that on the dowsing rod, suggesting that the author intended his work to satisfy this one inquiry.[24] Like Boyle, Vallemont adopted a mechanical natural philosophy, which he introduced after having described the rod and reviewed the Aymar case. All things in nature occurred by a singular mechanism, Vallemont explained, the movement of atoms in a void. This philosophy was older and therefore more credible than alchemical or Aristotelian theory.[25] It could explain all attraction or sympathy in nature, or the tendency for like things to approach and rest beside one another. If the atomic particles of a given mineral ore shared characteristics of the local vegetation, then a branch from that vegetation would respond to (hook up with) particles emanating from that ore.[26] This was the very same mechanism by which a dog caught the scent of a rabbit, the pollen of a plant reached its proper receptacle, and mercury attached itself to

---

[22] Heilbron, *Electricity in the 17th and 18th Centuries*, 222.

[23] "Leibniz an Wilhelm Ernst Tentzel: Clausthal, 26. Januar 1694." Leibniz, Band X, 224–226.

[24] Andreas Otto, tr., *Der heimliche und unerforschliche Natur-Kündiger, oder, Accurate Beschreibung von der Wünschel-Ruthe* (Nürnberg, 1694).

[25] Otto, *Accurate Beschreibung*, 52.

[26] "Man hat deren Wurckungen eben auch der Sympathie, so zwischen denen Metallen und einigen Pflantzen ist zugeschrieben, weiter kunte man nicht vorbringen." Otto, *Accurate Beschreibung*, 61.

gold.[27] Dowsing was no less natural than these common experiences, an instance of like particles that simply hooked together.

But why would a dowsing rod dip for a human being? Vallemont applied the Cartesian theory to the Aymar case, not by explaining the movement of his rod per se, but by claiming that the murderous matter about which Garnier spoke was a trail of sweat particles left by the murderer, like pollen strewn through the air. The pores on Aymar's skin captured the particles of the culprit, pulling the dowser's arms toward the source, as recognized by the dipping motion of the rod.[28]

These arguments occupied the first fourteen chapters of the text, after which Vallemont addressed dowsing practice among miners. "Among all those methods and ways through which one discovers mines, the very best is that which happens with the dowsing rod."[29] Vallemont proceeded to list standard treasure hunting beliefs, citing Agricola almost verbatim, and repeating a distinction the old mining-town physician had made between "skill" and "accident" in the discovery of mineral ore.[30] Accidental or naturally occurring events included the damage done by heavy rains, strong winds, lightning, earthquakes, and forest fires. Agricola included discoveries made by the farmer's plow. Mining folklore is full of such occurrences, which Vallemont presented as fact: miners founded Freiberg when, once upon a time, travelers from Bohemia chanced upon mineral ore after a rainstorm; and they settled Rammelsberg in the Harz Mountains after a horse's hoof upturned lead ore at Goslar. Agricola too recounted these and other stories as facts in *De veteribus et novis metallis* (1546), as did chroniclers such as Petrus Albinus in *Meißnische Bergk Chronica* (1590).

Discoveries resulting from human skill were preferable. "Rules acquired through long experience," said Vallemont, had greater value than the stories of accidental discovery.[31] Echoing Agricola again, he included tracking pieces of mineral rock embedded in the topsoil, as had happened, for example, at Kuttenberg in Bohemia, when a monk placed his habit (*Kutte*) over an exposed mineral fragment (founding

---

[27] Otto, *Accurate Beschreibung*, 97.

[28] "Die Atomi, so durch die *transpirationem insensibilem* deren Mördern zu Lyon in die Lufft zerstreuet, haben sich gar leicht durch die *respirationem insensibilem* in den Mann mit der Wünschelruthe ziehen können." Otto, *Accurate Beschreibung*, 193.

[29] Otto, *Accurate Beschreibung*, 376.

[30] Georg Agricola, *Zwölf Bücher*, 22.

[31] Otto, *Accurate Beschreibung*, 378.

Kuttenberg). The skilled digger also recognized the surface-level effects of mineral heats and vapors, such as evaporated frost, unusual characteristics of the foliage and vegetation, and the presence of certain flowers and plants. Vallemont listed juniper, ivy, figs, spruce, and other plants with "tips and thorns." The "experienced miner" (*Bergverständiger*) recognized the effects of mineral vapor. Vallemont was echoing the information Mathesius, Agricola, and more contemporary mining books provided.[32]

Vallemont then went beyond Agricola by distinguishing dowsing from all other methods as the most legitimate and reliable means of locating mineral ore. Vallemont claimed, again drawing on the experimental philosophy of Boyle, that other treasure-hunting methods were insufficient, because they did not provide specific information about the quality and depth of ore.[33] Vallemont was referring to Numbers 28 and 29 of the 'Inquiries,' in which the author raised this very issue. Contrary to Agricola, who upheld the so-called 'natural' signs of ore over use of the dowsing rod, Vallemont was confident that only the dowsing rod, properly understood, met these stringent requirements. He returned again to the corpuscular theory and claimed that since 'like attracts like,' a rod of gold would recognize the atomic particles of gold, one of silver would recognize silver-bearing ore, and so forth. The proof was in the results for Vallemont, who included a list of some 150 mines discovered in France by a famous dowser (the Baroness Martine de Bertereau): "A mile from Lorde, a silver mine. Near the towns of Pech and Verdun, three lead mines, a copper and an iron mine."[34]

Dowsing and the World Spirit

Vallemont's text served to publicize the strange Aymar case across Europe, and the corpuscular theory more broadly, as it purported to explain any number of mysterious phenomena. At Halle, the independent author and university librarian (*Auktionator*), Johann Gottfried Zeidler (1655–1711), similarly defended dowsers, although he rejected the fashionable French theory. Halle, newly acquired by Friedrich Wilhelm I of Brandenburg-Prussia, was well known as the

[32] Otto, *Accurate Beschreibung*, 381.
[33] Otto, *Accurate Beschreibung*, 384.
[34] Otto, *Accurate Beschreibung*, 385–7.

center of the pietist movement. Pietism, a highly anti-dogmatic branch of Lutheranism, stressed personal religious experience and social works. The University of Leipzig had banned the pietists, and the Elector of Saxony had exiled their leaders in 1690. Friedrich Wilhelm, who would make pietism a state religion, founded the University of Halle in 1694 partly as a refuge for these exiles. But the university would also employ some of the leading figures of the early German Enlightenment and anti-Cartesian philosophy more broadly, including Christian Thomasius, Christian Wolff, Friedrich Hoffmann, and Georg Ernst Stahl. He in particular was instrumental in developing a veritable alternative to Cartesian philosophy that historians have come to call 'vitalism.'

Peter Reill argues that the rejection of the mechanical philosophy in the eighteenth century was not limited to the Romantic science (*Naturphilosophie*) of Friedrich Schelling, Johann Christian Reil, or Lorenz Oken at century's end. There was another Enlightenment tradition. Two of its early developers were the natural historian in France, the Comte de Buffon, and the chemist, Stahl. The historiography has tended to identify Stahl exclusively with the phlogiston theory of combustion, and accordingly, to privilege his 'revolutionary' successor in chemistry, Antoine Lavoisier. But Reill recovers the distinctiveness of Stahlian chemistry and sets it into the context of a larger anti-Cartesian movement that came to influence natural history, biology, chemistry, and medicine in Germany.[35]

Buffon and Stahl attacked the mechanical philosophy for its failure to explain living matter, or its rash rejection of the fundamental assumption (both Aristotelian and hermetic) that nature possessed a principle of animation. The new philosophy of Boyle or Descartes had reduced all matter to particles extended through space, subject to simple laws of motion, and devoid of any qualities or internal drives. Descartes drew a sharp distinction between mind and body that made automatons out of animals and most human activity. Many vitalists rejected his definition of matter and motion, and sought rather to unify mind and body by reviving the doctrine of vital forces in nature. Drawing in part on the alchemical tradition (Paracelsus, Jan Baptista van Helmont, among others), the vitalists saw elective affinities,

[35] Peter Hanns Reill, *Vitalizing Nature in the Enlightenment* (Berkeley, CA: University of California Press, 2005).

sympathetic attractions, and various drives in nature. Rather than the simple cause/effect relationships that the mechanists preferred, the vitalists saw interconnections and wholeness in motion, in which all parts interacted simultaneously with all others, and vital forces operated toward final ends. Internal drives were the causes of motion, perception, and development, not cause/effect relations.[36] Much of vitalism would indeed resemble later *Naturphilosophie* of the Romantic scientists, but they were far more speculative and bolder in their claims than Enlightenment vitalists, as Reill defines them.[37]

The vitalists differed markedly among themselves, they were not necessarily *anti*-mechanical, and there were French vitalists as well as German mechanists.[38] Leibniz may have done as much to advance vitalism as the mechanical philosophy. But what distinguished vitalism, irrespective of who best represented the movement, was a commitment to explaining life and motion in terms of a non-mechanical spirit, force, or virtue. We should stress that Stahl's student, Johann Friedrich Henckel, who we met in Chapter Four, would introduce vitalist chemistry to Freiberg. In *De mediorum Chymicorum appropriatione* of 1727, for example, Henckel spoke of the affinity between chemicals that brought them into reaction.[39]

It is no surprise that a Halle scholar of the first generation such as Zeidler would reject the materialist tendencies of the French, and posit an alternative explanation for Aymar's feat. We cannot identify Zeidler with Pietism exactly—although he was certainly non-orthodox and worked for the University of Halle—but can identify his natural philosophy with the vitalist tradition.[40] The *Pantomysterium* (1700)

---

[36] Reill, *Vitalizing Nature*, 6–13, 71–77.

[37] Reill, *Vitalizing Nature*, Chapter Five.

[38] On vitalism in the French context, see Elizabeth Williams, *A Cultural History of Medical Vitalism in Enlightenment Montpellier* (Aldershot, Hants and Burlington, VT: Ashgate, 2003). On vitalism and early embryology, see Shirley A Roe, *Matter, Life and Generation: Eighteenth-Century Embryology and the Haller-Wolff Debate* (Cambridge, NY: Cambridge University Press, 1981). On vitalism and Romantic science, see Reill, "Science and the Construction of the Cultural Sciences in Late Enlightenment Germany: The Case of Wilhelm von Humboldt," *History and Theory* 33: 3 (Oct. 1994), pp. 345–366.

[39] The Russian chemist Michael W. Lomonossow remarked that Henckel rejected Cartesian philosophy: "[E]r verachte alle vernünftige Philosophie; das besagt, daß er die mit Descartes einsetzende philosophische Richtung ablehnte." Walther Herrmann, *Bergrat Henckel: Ein Wegbereiter der Bergakademie* (Berlin: Akademie Verlag, 1962), 155.

[40] The little we know about Zeidler is provided in Cornelia Niekus Moore, "Susanna Elisabeth Zeidler (1657–1706), Leben und literarisches Wirken einer Dichterin am

exposed Zeidler's non-traditional religious beliefs and developed a vitalist theory of dowsing. He provided a history of the practice, translation of letters between Malebranche and the Jesuit, Pierre Lebrun, concerning the rod's supposed atomic workings, Zeidler own response to this theory, and an overview of his vitalist system. The central theme to which Zeidler repeatedly turned throughout was his rejection of university or scholastic tradition, and especially his refutation of Cartesian philosophy. For Zeidler, "scholars" included all such men. Aside from the latest theory from France, the scholars tended to discount the dowsing rod, and did so for four main reasons: preoccupation with identifying superstitions, inability to think in non-scholarly terms (to see "beyond their glasses"), unwarranted distrust of popular and artisanal knowledge—"one does not put the discoveries of farmers, miners, and craftsmen in the *Acta Eruditorum*"—and because such men assumed that everything they did not understand was somehow unnatural.[41] Zeidler censured the Cartesians, not for rejecting dowsing, but rather for presuming to explain all things, however mysterious or spiritual, in mechanical terms. He ridiculed their fixation on invisible particles and 'effluvia,' and their inexperience with the phenomena they claimed to understand. "How bold" of these men to take on dowsing. They were committed "merely to experiment," rather than to true experience.[42] Zeidler attacked the Abbot of Vallemont in particular as a "mere materialist and atomist" who misrepresented the dowsing rod because he had little personal experience of it.[43]

Zeidler proposed an alternative dowsing theory based largely on the thought of his famous contemporary at Halle, Christian Thomasius, some of whose work Zeidler had translated into German, and who would preface Zeidler's work on dowsing. Thomasius was the author of many polemical tracts and lectures at Leipzig that generated great popularity for the author, but also scorn. His popular-science journal, the *Monatsgespräche*, which included satirical and social-critical pieces, made Thomasius into a pariah in Saxony. In 1690, after a formal charge of atheism had been levied against him, Leipzig canceled his

---

Ende des siebzehnten Jahrhunderts," in Cornelia Moore, ed., *Susanna Elisabeth Zeidler: Jungferlicher Zeitvertreiber* (Bern: Peter Lang, 2000).

[41] Johann Gottfried Zeidler, *Pantomysterium, oder das neue vom Jahre in der Wündschelruthe als einem allgemeinen Werckzeuge menschlicher verborgenen Wissenschafft* (Magdeburg, 1700), 112.

[42] Zeidler, *Pantomysterium*, 200.

[43] Zeidler, *Pantomysterium*, 109.

appointment, and Thomasius was forced to leave. A local preacher publicly burned Thomasius's work, as he fled to Berlin. With the support and protection of a more tolerant Elector, he migrated to Halle, where Thomasius made the acquaintance of the leading Pietist, August Hermann Francke. But Thomasius was committed to certain ideals of the early Enlightenment that he would have considered at odds with pietist enthusiasm. For example, he sought in his written work to distinguish religion from law and morality, and he became an outspoken opponent of witch-hunting. In 1710, notwithstanding ongoing conflict with the pietists, Thomasius became the rector of the University of Halle.[44]

Zeidler would draw especially on Thomasius's *Versuch von dem Wesen des Geistes* (1699), published just one year prior to Zeidler's *Pantomysterium*. The *Versuch* presents a thoroughly vitalist natural philosophy formulated in opposition to Cartesian theory.[45] Thomasius attacked the materialists for failing to explain motion and life. If the universe were filled with matter (a plenum), with no empty space, as the new philosophy maintained, there would be no cause for motion, attraction, pressure, and other physical phenomena. To account for these things, Thomasius supposed that a non-material entity must occupy the spaces between particles. This 'unmoved mover' is the spirit or soul (*Geist*) of the universe and the cause of all motion. As he analyzed this spirit more closely, Thomasius came to distinguish three forms: pure spirit (God), light, and air. The second and third forms were spiritual rather than material entities. Both light and air carried the universal spirit throughout the universe, as the medium of spirit. Thomasius discussed the classical four elements as well, but he identified both fire and air as immaterial agents operating in unison with the world spirit. Spirit mixed with water and earth in various degrees, and

---

[44] Josef Rattner and Gerhard Danzer, "Christian Thomasius—der Erfinder der deutschen Aufklärung und Bekämpfer des Hexenwahns," in *Philosophie im 17. Jahrhundert: die Entdeckung von Vernunft und Natur im Geistesleben Europas* (Würzburg: Königshausen & Neumann, 2005), 187–202. See also W. Schmidt, *Ein vergessener Rebell—Leben und Wirken des Christian Thomasius* (Munich, 1995).

[45] *Versuch von Wesen des Geistes Oder Grund-Lehren, so wohl zur natürlichen Wissenschafft als der Sitten-Lehre. In welchen gezeiget wird, dass Licht und Lufft ein geistiges Wesen sey, und alle Cörper aus Materie und Geist bestehen, auch in der gantzen Natur eine anziehende Krafft, in dem Menschen aber ein zweyfacher gutter und böser Geist sey* (Halle, 1699). On the *Versuch* in context of Thomasius's relation to the pietist movement, see Martin Gierl, *Pietismus und Aufklärung* (Göttingen: Vandenhoeck und Ruprecht, 1997), 437–458.

the more endowed with spirit an object was, the greater the extent to which it partook of universal spirit. Animals, plants, and minerals all possessed the living principle of universal spirit via light (especially by the sun) and air, though only mankind possessed a high degree of light and air vis-à-vis earth and water, and the capacity to think and behave in some agreement with pure spirit. Thomasius charted the complex relationships he saw between the universal spirit, the Creation, and mankind.[46]

Universal spirit might appear to be divided into three (God, light, air), but this Trinity was actually unified, and all bodies that partook of universal spirit were united in spirit. This was the reason that Thomasius believed that minerals and stones had healing virtues for man, that the human body contained plant-like properties like seeds and vessels, and that it resembled the animal body in more obvious ways. The universal spirit promoted an archetypical pattern of generation and life cycle in all these living things. The spirit of man (which Thomasius subdivided further into thinking, feeling, and nourishing functions), moreover, was the cause of fellow feeling between men, and other psychological phenomena, as we might call them today. Thomasius refrained from granting the human mind any magical power per se, but he did allow for profound insight into the universe: certain abstractions and concepts that, by tapping a more universal knowledge, might provide insight into past and future events, and even an awareness of hidden or distant happenings.[47]

Zeidler did not have to rework his colleague's philosophy to apply it to dowsing. Parts of the *Pantomysterium* were in fact recapitulations of Thomasius's theories. In a section called, *Einleitung zur natürlichen Philosophie*, Zeidler, in 127 theses, laid out his cosmology and natural philosophy, much as Thomasius had done in 195 theses in the *Versuch*. Zeidler began with the sun, which he identified as the source of all life, motion, and change on earth. He quickly identified the sun's life-giving power as a spirit or soul (*Geist*) that traveled along rays of light. Light, just as Thomasius had said, was a medium for spirit rather than a true

[46] Thomasius, *Versuch*, 98.

[47] "Die Krafft des Gehirns ist von denen Thieren unterschieden, dass der Mensch, vermittelst der Gedancken, nicht allein die gegenwärtigen Dinge in seiner Einbildung fassen, sondern auch vermittelst der abstractionen und Begriff derer Universalium oder ideen auch von abwesenden, vergangenen und zukünfftigen Dingen dencken kan." Thomasius, *Versuch*, 185.

element; nor was fire a real element, but rather a manifestation of light. Absent this spirit in matter, the principle cause of motion, fire was merely dead material.[48] Zeidler, who had studied mathematics and astronomy before moving to Halle, devoted more space to the sun than had his colleague: the spirit of the sun, which revolved upon its own axis from the center of the universe, held the planets in their orbits. As Zeidler developed his theory of the world spirit, he pitted it against the mechanical philosophy, and in particular the cosmology that Descartes proposed in *Le Monde* (~1633).

Zeidler, drawing on Thomasius, was certain that the spirit of man could access a more universal reservoir of knowledge, but Zeidler used dowsing to demonstrate the point. The spirit of the dowser participated in the larger world spirit, which caused motion in a properly held rod. The practitioner possessed an idea of mineral ore that actually led to its discovery. The spirit of the ore (*Sachgeist*), or of any number of hidden objects for that matter, became bound to the dowser's own spirit, which brought it to his imagination. The process was a function, "not only of the hand, but of our mind."[49] The skilled dowser could draw from the world spirit to locate mineral veins, springs, hidden treasures, and lost items; track fleeing criminals, lost persons, runaway livestock, and wild game; determine where somebody had been; measure distances; even distinguish between true and false statements.[50] The power of the dowser's mind when it tapped the universal spirit was not unlike the rays of the sun when concentrated by a magnifying glass.

Thomasius did not see this easy parallel between his philosophy and dowsing, though he nonetheless wrote Zeidler's preface. Thomasius sympathized with this attempt to ground dowsing practice on a new foundation that would avoid both the materialism of the Cartesians and the dogmatism of the religious establishment. While he ridiculed any number of supposed successes with the dowsing rod that Zeidler's text reported as fact, Thomasius remained open-minded and admired the larger purpose. He did not believe that the dowsing rod could be used to determine pregnancy, the truth-value of statements, or even whether treasure lay in a castle wall, among other examples Zeidler provided. Thomasius allowed only that dowsing had a role in locating

[48] Zeidler, *Pantomysterium*, 92.
[49] Zeidler, *Pantomysterium*, 515.
[50] Zeidler, *Pantomysterium*, 547.

mineral veins, water sources, and, indeed, the culprit of a murder. The rod made the discovery of these things, "at least more likely."[51]

Thomasius took issue, not so much with dowsing per se, or with Zeidler's overindulgence with the *Vesuch*, as with French philosophers and dogmatic theologians. "If I were to embark on a disputation with a Cartesian, Gassendian, or other new philosopher, and show them that everything they claim to prove and solve with their teachings is pure rubbish, I would simply ask that they explain to me clearly the phenomena of the human soul, and so long as this is not performed, that they keep quiet with their Physic."[52] Thomasius could list any number of examples of complex phenomena the materialist philosophy was unable to explain, such as the spread of yawning or laughing among people, or the experience of falling in love. He numbered dowsing among these examples, and so he supported Zeidler for attempting an explanation that took account of the activity of the human spirit, rather than reduced everything to corpuscles in motion. As for the religious authorities, Thomasius was furious that so famous a scholar and theologian as Malebranche would rush to the charge of diabolism rather than examine the phenomenon more critically. Did he not see how fervently poor miners prayed to God? How they used the dowsing rod with no sorcery and strange incantations, but rather insisted on its natural functioning?[53] The hatred that Thomasius and Zeidler shared for the Cartesians, and their commitment to a vitalist philosophy, brought these Halle scholars together, notwithstanding their differences concerning the role of the world spirit in dowsing practice.

Certainly, not all natural philosophers in these decades were willing to dignify dowsing with new science, whether mechanical or vitalist. Leibniz had a greater commitment (and commission) to refashioning the mining industry than Boyle, Descartes, Thomasius, and most other contemporary philosophers did; and prospecting beliefs, among them dowsing, served only to impede the progress and change Leibniz wished to spearhead. In his writings on mining and earth science we detect a greater 'withdrawal' of an elite from popular mining culture (P. Burke), a clearer distinction between science and superstition, and a louder call for professional training than in Boyle, Vallemont, or

[51] Zeidler, *Pantomysterium*, Preface, A6.
[52] Zeidler, *Pantomysterium*, Preface, A3.
[53] Zeidler, *Pantomysterium*, Preface, B2.

Zeidler. In this sense, Leibniz presaged the high Enlightenment and Mining Academy, to which we turn shortly.

The Philosopher of the Harz Mountains

Leibniz was an advisor and librarian to the Guelf dukes of Braunschweig-Lüneburg at Hanover from 1676. His position and location, far removed from more important centers of intellectual life, never afforded Leibniz the prestige and freedom that he felt he deserved. Leibniz's sense of isolation and struggle to secure a higher status and income at court surely motivated his grandiose schemes and projects for the Guelf dukes, plans for Harz mining among them, beyond his central commission to write the Guelf House history.

But a disjoint between theory and practice, or between "dreams and reality," had already come to characterize Leibniz's career.[54] He tried unsuccessfully as early as 1669 to found a German version of the famous western societies that gave him membership, but with which corresponding was always difficult. The Academy of Natural Curiosities (*Academia Naturae Curiosorum*), or Leopoldina (after Leopold I), organized in 1652 at Halle, focused more on medicine and botany than on physical and mathematical sciences. It also lacked the same coherence as the western societies: it was a private institution with limited funding, and included only twenty members after ten years. In 1676, Leibniz formulated further plans for a German academy, including a list of fifty prospective members, but again, in vain.[55] The university remained the major site of natural inquiry in Germany before the Berlin-Brandenburg Society of Sciences (later Prussian Academy) opened in 1700 with Leibniz as president. It did not publish before 1710, however, and its heyday was after mid-century, when Maupertius, Euler, and Lagrange held high positions.

Leibniz's plans for improving Harz mining were one among many failed projects he proposed during his stay in Hanover. Altogether, he

[54] Maria Rosa Antognazza, *Leibniz: An Intellectual Biography* (Cambridge, NY: Cambridge University Press, 2009), 195.

[55] Conrad Grau, *Die Preußische Akademie der Wissenschaften zu Berlin: Eine deutsche Gelehrtengesellschaft in drei Jahrhunderten* (Berlin: Spektrum Akademischer Verlag, 1993), 24–7. See also Ayval Ramati, "Harmony at a Distance: Leibniz's Scientific Academies," *Isis* 87 (1996), 430–452, and Fritz Hartmann and Rudolf Vierhaus, eds., *Der Akademiegedanke im 17. und 18. Jahrhundert* (Bremen: Jacobi Verlag, 1977).

spent about three years working in the mining towns, and between 1680 and 1685 alone he visited some 31 times.[56] He sketched his grandest plans in a 1682 work entitled, *Denkschrift betreffend die allgemeine Verbesserung des Bergbaues im Harz* (Memorandum Concerning General Improvement of the Harz Mines). In this essay, Leibniz described an ideal of rationalized bureaucratic control. It contained the outline for a more complete mining text, and proposals for better management of the mines and application of new technologies, especially windmills (rather than waterwheels) for draining water. As Hamm describes, Leibniz's *Denkschrift* combined the pursuit of better knowledge of the earth with a vision of central management.[57] Not surprisingly, Harz miners and officials seemed to have disliked the new philosopher and court librarian, and self-designated mining technician, and Leibniz in turn expressed negative comments about them.[58]

Leibniz divided the first part, a "complete account" (*vollkommene Nachricht*) of mining, into *Universalia* and *Singularia*, or that which pertained to mines in general and that which varied from place to place. Beginning with general knowledge, Leibniz, like Agricola and other authorities on mining, began with the first step in any mining enterprise, digging.[59] Leibniz called for an account of the indications or signs (*Anzeigungen*) of underground ore, and of the science (*Wissenschaft*) of stones, by which he meant the knowledge of how mineral fragments confirmed the layout of veins and quality of ore.

---

[56] Günter B.L. Fettweis, *Zur Geschichte und Bedeutung von Bergbau und Bergbauwissenschaften: 21 Texte eines Professors für Bergbaukunde zur Entwicklung des Montanwesens in Europa und speziell in Österreich* (Vienna: Verlag der Österreichischen Akademie der Wissenschaften, 2004), 195. On Leibniz's plans for mining, and his mineralogical and geological writings, see also Roger Ariew, "Leibniz's Protogaea," in *Leibniz: Tradition und Aktualität: V. Internationaler Leibniz-Kongress*, Ingrid Marchlewitz, ed. (Hannover, 1988). Claudine Cohen, "Leibniz's Protogaea: Patronage, Mining, and Evidence for a History of the Earth," in *Proof and Persuasion: Essays on Authority, Objectivity, and Evidence*, Suzanne Marchand and Elizabeth Lunbeck, ed. (Turnhout: Brepols, 1996). Hamm, "Knowledge from Underground." Claudine Cohen and André Wakefield, eds., *Protogaea: Gottfried Wilhelm Leibniz* (Chicago: The University of Chicago Press, 2008).

[57] Hamm, "Knowledge from Underground," 84.

[58] "It was not long before he had angered and alienated many of the local miners and officials there. He complained that they were obstructing his plans, and even sabotaging his work; they complained that his work was impractical and would waste money." Cohen and Wakefield, *Protogaea*, xvii.

[59] "Derowegen so müste man anfangen von den Anzeigungen deren man sich alhier bedienet einen verborgenen Orth zu finden, alda vermuthlich Erz anzutreffen." *Gottfried Wilhelm Leibniz: Allgemeiner Politischer und Historischer Briefwechsel* (Leipzig: K.F. Koehler Verlag, 1938), 151.

He proposed to develop this science in a subsequent work. Here in the *Denkschrift*, Leibniz referenced the 1666 'Articles of Inquiries Touching Mines' of the Royal Society, noting only that "some are helpful, others vastly unhelpful."[60] Leibniz would not merely report prospecting tradition, however curious, as the 'Articles' had done, but replace it with a more rigorous science.

In Leibniz's prospective text, after an explication of stones and earths, there would follow an overview of mine surveying, mining law, machine technology (ventilation, hydraulics, and hauling), sorting, and smelting. In this last section, Leibniz wished to build on the famous 1574 assaying manual of Lazarus Ercker, one-time mint master at Goslar in the Harz Mountains. Turning to the *Singularia*, or particular knowledge of mining, this involved a *Topographiam Subterraneum*, or complete account of the waters, forests, and earths of different mining regions. It would include cross-sectional charts, "as if the eye were suspended in air and the mountain were transparent."[61] Detailed knowledge of this sort should ground the decision making of managers who risked tunneling merely because of prior "luck," or discontinued out of "ungrounded fear."[62]

In the second part, on management, Leibniz extended his plans for total knowledge of the mines to control over production. He called for weekly and monthly reports on conditions and changes at the mines, blind visits, tighter oversight by managers and trusties (*Geschworene*), more efficient use of water and wind, and more precise surveying instruments and compasses. Other proposals included providing prospective officials and foremen with better instruction, the means to visit foreign mines—"normally they do not go any further than Freiberg"—and patterning the relationship between foreman and worker on that of master and journeyman. Advancement to foreman should be based on merit and skill (including some sort of masterpiece) rather than favor and friendship.[63] Tighter control over the workers themselves would require more accurate time-keeping to ensure, for example, that one man did not sign in and collect wages at two mines simultaneously. Leibniz even called for more meticulous accounting of the daily activities of workers to determine how diligent

[60] Leibniz, *Briefwechsel*, 152.
[61] Leibniz, *Briefwechsel*, 159.
[62] Leibniz, *Briefwechsel*, 159.
[63] Leibniz, *Briefwechsel*, 163.

they were, whether they deserved a full wage, and whether they should be kept, promoted, released, or punished.[64]

Leibniz met resistance with the Mining Office over funding, as well as technical difficulties with his windmills, and in 1685, the Duke requested that Leibniz discontinue the work.[65] His plans for mining were never fully implemented, and remain an unfulfilled ideal of knowledge and rational administration for the cameral state. Still, Leibniz visited Freiberg in 1680 and again in 1688, during a tour of Bohemian and Saxon mining towns, including Annaberg, Marienberg, and Schneeberg. He met with major mining officials and probably with the chief overseer of Saxon mining, Abraham Schönberg.[66] Leibniz's reforming zeal surely found resonance at Freiberg.

If Leibniz imagined disciplining the bodies (M. Foucault) of workers and officials in the *Denkschrift*, he discredited their beliefs in another work on earth science. The German counterpart to the *Philosophical Transactions* was the *Acta Eruditorum*, printed in Leipzig beginning in 1683. It published Boyle, Leeuwenhoek, Bernoulli, Pascal, and Descartes, among other luminaries, and served to position Germany in the emerging Republic of Letters. Leibniz's *Protogaea*, likely composed while he was working in the Harz Mountains, is rich in observations of gullible (*leichtgläubige*) workers. Leibniz intended it to be the introduction to the commissioned history of the house of Braunschweig-Lüneburg, but it quickly expanded beyond the stated purpose, and afforded the author an opportunity to pursue his broader interests. *Protogaea* first appeared in summary form in the *Acta Eruditorum* in 1693, and it circulated in manuscript until the 1749 German edition.

With the goal of founding a new science of 'natural geography' (*geographiam naturalem*), Leibniz described a mechanical cosmogony of the earth from a primitive molten state to its present form. Like other contemporary theorists of the earth, Leibniz took Creation and the Deluge seriously, but he was no literalist, and allowed for more than one period of flooding. Long phases of cooling, evaporation, and condensation created subterranean crusts, waters, and vapors that, in turn,

---

[64] Leibniz, *Briefwechsel*, 164.

[65] Ramati, "Harmony at a Distance," 440.

[66] Jobst and Schellhas, *Abraham von Schönberg—Leben und Werk: Die Wiederbelebung des erzgebirgisches Bergbaus nach dem Dreißighährigen Krieg durch Oberberghauptmnan Abraham von Schönberg* (Leipzig: Deutsche Verlag für Grundstoffindustrie, 1994), 46.

caused countless earthquakes, floods, and volcanoes that ultimately gave the earth's surface its present-day landscape. Indebted chiefly to the works of Descartes, Boyle, Steno, and even Agricola, Leibniz also drew on metallurgy and alchemy to describe the 'crystallizations,' 'sublimations,' and 'precipitations' of these processes. Contrary to Thomas Burnet's theory that the earth was in a state of decay from a perfect edenic form, Leibniz contended that it had achieved relative equilibrium, though it was still subject on a smaller scale to the same geological phenomena of prior ages.

In addition to this account of the formation of the earth, the science of natural geography would also involve an exhaustive description of the layout of strata and minerals, and rules for the location of ore.[67] A great obstacle in this respect, according to Leibniz, was popular digging knowledge, and he ridiculed prospecting beliefs and practices: "I do not doubt that someday, through careful observations, it will be possible to construct principles that can yield rules governing what is hidden under the earth, like metal ores; and I believe these rules can be far better than those generally praised on weak grounds, more through tradition than success, and above all because the miners of some regions apply them."[68] It is instructive to compare these words with those of Nicolaus Steno in the 1669 *Prodromus*, to which Leibniz refers repeatedly in these sections. Steno cited dowsing in particular and associated it with a certain form of sorcery practiced in China:

> The minute and all but superstitious divisions of veins utilized by miners are built on the slightest foundations, nay clearly on no foundation at all; so that the divination of an abundance of minerals by means of roots and twigs is as dubious with regard to metals as the opinion of certain Chinese is ridiculous with regard to the head and tail of the dragon that they use in discovering a favourable position for burial places on mountains.[69]

Leibniz also took issue with the assumption that minerals grew like plants under the earth, complete with branches and stems. Leibniz offered an explanation for this widespread error: he claimed that miners considered the lines and markings laid by the mine surveyor on the

---

[67] Hamm, "Knowledge from Underground," 82.

[68] Cohen and Wakefield, *Protogaea*, 25.

[69] Nicolaus Steno, *The Prodromus on Solids* (Odense: Odense University Press, 1969), 171. Steno in fact joined the Hanover court in 1677. Unfortunately for Leibniz, who felt isolated, the Dane had by then largely abandoned his natural philosophical studies. Antognazza, *Leibniz*, 202.

earth's surface to be plantlike in form, and deduced that the mineral veins must be plantlike in nature. Without doubt, he explained, God originally created the soil much like he did plants and animals, but over the centuries great geological phenomena had obscured the original patterns, such that they had become almost unrecognizable.[70] And while men in mining believed that the Rammelsberg tunnels (near Goslar) had actually shrunk, as if regenerating over time, in fact waters carried sediment through the tunnels and deposited it at predictable locations.[71] As for the alchemists: they could not mimic or hasten nature's handywork, natural heat was far greater than that which man could reproduce in the furnace, and man only rarely observed nature in the process of forming mineral earth, in any case.[72] Sands that had once contained gold would not regenerate noble metal after prolonged exposure to the sun.[73] Leibniz accordingly judged the countless claims of alchemists and sympathetic miners to be largely unsubstantiated, citing again their great gullibility.

Leibniz also commented on mining spirits and the dowsing rod. His remarks on the spirits were particularly inventive: supposed sightings, such as the finger-long dwarf dressed in digger's uniform somebody witnessed carrying a full basket of metal, merely demonstrated the power of imagination (*Einbildungskraft*). Sometimes we project onto nature what we worship daily or observe in other contexts.[74] The capacity of the human mind to self-deceive was a favorite explanation for enlightened authors a century later. In his *Book on Superstition* (1791), Heinrich Ludwig Fischer claimed that "experienced miners" (*Bergverständige*) denied ever having witnessed a mining spirit, but that others, half-inebriated and surrounded by dust and fumes, saw human-like images.[75] Always ahead of his time, Leibniz scorned the writings of contemporary philosophers who spread this sort of

---

[70] Cohen and Wakefield, *Protogaea*, 25.

[71] Cohen and Wakefield, *Protogaea*, 29.

[72] Cohen and Wakefield, *Protogaea*, 27.

[73] Cohen and Wakefield, *Protogaea*, 27.

[74] Cohen and Wakefield, *Protogaea*, 75.

[75] "...aber die Bergleute pflegten vor ihrer Einfahrt gewöhnlich hitziges Getränk zu sich zu nehmen, wodurch ihre Einbildungskraft gespannt, und ihre Vorstellungen verwirrt würden; da es dann leicht möglich sey, daß sie etwas sehen, was doch gar nicht da sey. Unter der Erde in unterirdischen Höhlen und Bergwerken sammeln sich überdem leicht Dünste, die halbberauschten Bergleuten (denn nur solche sehen das Männchen), besonders wenn ihre Einbildungskraft wirkt, allerdings in gewissen Gestalten erscheinen können." Quoted in Heilfurth, *Bergbau und Bergmann*, 166.

nonsense, including the "fairy tales" (*Märchen*) of Johann Joachim Becher and Athanasius Kircher, who claimed, like simple commoners, that stones and fossils resembled plants, animals, stars, and even classical and biblical figures like Apollo and Moses. Even the Pope and Luther were supposedly etched in stone at Eisleben.[76] Not surprisingly, Leibniz gave the dowsing rod the same treatment as Steno did, associating it with mining spirits and other superstitions that only the gullible believed, despite the fact that the dowser, when blindfolded, could not discover even the largest and best-known veins with his twitching stick.[77]

Mining spirits, mineral trees, dowsing rods—to the man who challenged Newton for priority of the calculus and imagined a total revamp of Harz mining, these were the equivalent of cracking an egg in water to read the future, or spilling oil on a board to interpret the lines. The practices of "superstitious little women" (*abergläubische Weibleine*) had no business in proper philosophical texts.[78] Given the widespread use of the dowsing rod among men, and miners and mining officials at that, Leibniz's association of the practice with the popular magic of women is significant. His vision of total control over mining production and knowledge was a masculine ideal, and the superstitions and traditions of diggers were a feminine principle Leibniz wished to extricate from mining. Historians have noted how markedly gendered the language of earth science specialists was, from Leibniz to Adam Sedgwick.[79]

Leibniz was not alone in gendering the concepts of 'science' and 'superstition,' though others would revive the masculine ethic behind dowsing. His contemporary, the independent scholar and sensationalist writer in Leipzig, Johann Praetorius (1630–1680), adopted the

---

[76] Cohen and Wakefield, *Protogaea*, 73.

[77] "One blindly believes the stories of people who have accustomed themselves, through their own credulity and that of others, to deceiving and being deceived—people who see homunculi and Plutonic monks nearby while they work underground, and who search the earth for hidden treasures with divining rods, even though, when you bind up their eyes, they do not find the largest and most famous veins through the sign of the quivering rod." Cohen and Wakefield, *Protogaea*, 27. By contrast, Becher summarized Thölde's theory of mineral vapors and dowsing rods in *Natur-Kündigung der Metallen* (Frankfurt, 1660), 85–100.

[78] Cohen and Wakefield, *Protogaea*, 75.

[79] Hamm, "Knowledge from Underground," 93. Michael Shortland, "Darkness Visible: Underground Culture in the Golden Age of Geology," *History of Science* 32 (1994), 38–43.

same intellectual and gendered divide. Praetorius ridiculed numerous "superstitions of the common man" in his entertaining (and aptly titled) *The Philosophy of Women* (1662),[80] but he, unlike Leibniz, defended the dowsing rod as *Wissenschaft* shortly thereafter (1667). Recalling the application of dowsing in mine surveying, Praetorius emphasized the analogy between the lodestone (magnetite) and dowsing rod. Otherwise an entertaining review of treasure hunting stories, Praetorius's work assumed a more serious tone when it set the words of Valentine (Thölde), Agricola, Paracelsus, and Mathesius on dowsing against the arguments of the Wittenberg professor, Johann Sperling, who argued that the practice required an implicit pact with Satan. Dowsing was not witchcraft, contended Praetorius, but was grounded in the experience and testimony of miners. He upheld their honor and credibility: "Do not discount the high law of mining custom...why would you discount it and interpose your own authority?"[81] The "mining rod" did not require special incantations or other rituals, which would be "either witchcraft...or fraud," if not simple foolishness.[82] Praetorius acknowledged that there were superstitions in dowsing, especially in procuring the rod, but he outlined proper practice. It involved a hazel branch that had grown above mineral ore. Having imbibed the mineral's virtue through its roots, the rod showed a "natural desire" to return to the earth from which it came, responding, like Thölde had said, to the "breath" (*Anhauchen*) of metals.[83] If the rod was impotent for Leibniz, it was a powerful tool for uncovering hidden treasure for Praetorius.

From Republic to Academy

This chapter outlined some of the learned opinions concerning mining and dowsing in the decades prior to the Freiberg Academy. Mining

[80] Johann Praetorius, *Philosophia colus oder Vfylosovieh der Weiber, darinnen gleich hundert allerhand gewöhnliche Aberglauben des gemeinen Mannes lächerig wahr gemachet werden: die kurtze Zeit zu verlängern, und die lange Zeit zu vertreiben* (Leipzig, 1662). Wilde calls Praetorius the "first Saxon folklorist." Wilde, *Die Zauberei- und Hexenprozesse*, 449.

[81] "Verwirfft doch das hohe Recht den Bergwerckischen Gebrauch nicht; warumb wilt du ihn verwerffen, und deine schlechtere Authorität interponiren?" Johann Praetorius, *Gazophylaci Gaudium*, Preface.

[82] Praetorius, *Gazophylaci Gaudium*, Preface. On Sperling and witchcraft, see Chapter Two.

[83] Praetorius, *Gazophylaci Gaudium*, Preface.

was a major state interest, and natural philosophers and savants across Europe sought credible knowledge of minerals and the earth. After the Aymar case in France, some scholars elevated the rod to the forefront of contemporary philosophy, inciting a quick reaction from anti-Cartesians, particularly at Halle. Leibniz approached the issue from a different angle, and sought to implement overarching reforms in mining. He called for the re-education of officials and a higher level of bureaucratic control, and he gave short shrift to miner beliefs like dowsing, anathema to the robust masculine ideal he wished to realize for the Guelf. In all, scholars of the Scientific Revolution and early Enlightenment handed contemporary mining officials at Freiberg—men like Augustus Beyer and Johann Friedrich Henckel—a host of arguments, but no definitive statement or consensus view. The Mining Academy was conceived along lines Leibniz would have approved, but both teaching and mining practice required that professors and officials consult the *Bergwissenschaft* of old, and miner wisdom prevailed in the matter of prospecting. As for dowsing itself, it was not Vallemont's corpuscularism or Zeidler's vitalism that came to replace mineral vapor theory, but that synthesis of pre-modern magic and new science that was *Naturphilosophie*.

CHAPTER SIX

THE ELECTRIC ROD: DOWSING AND THE FREIBERG MINING ACADEMY

In 1780, the Göttingen Academy of Sciences sought to generate further research on an important topic by publicizing the following essay contest: 'How was mining conducted in prior centuries? Can we learn something of benefit to present-day mining and smelting by comparison?' The prize-offer remained open until 1783, when five contestants, among them Christian Hieronymus Lommer, submitted papers. Lommer was the first professor of mine engineering (*Bergbaukunde*) and mineralogy at the Freiberg Mining Academy in Saxony.[1] Though he did not win, the contest afforded Lommer an opportunity to showcase Freiberg, fast becoming an internationally recognized center of earth science, to his distinguished Göttingen colleagues, men such as Johann Friedrich Blumenbach.[2] Lommer's response also demonstrated how Freiberg scholars invoked the dowsing rod to distinguish Academy science from mining superstitions.

Lommer cited numerous examples of a mining culture he claimed was in decline. Contemporary mining science at Freiberg was, he said, grounded on reason and a combination of different theoretical and practical branches, including chemical mineralogy, natural history, physics, geology (*Lagerstättekunde*), mine surveying, mechanics, hydraulics, and hydrostatics. Lommer pitted these fields against mining

---

[1] *Göttingische Anzeigen von gelehrten Sachen*, nr. 202 (1783): 2027–8. The original offer was repeated in 1781 and 1782, before the Society awarded two of five contestants in 1783 the prize of fifty ducats, to be shared. Christian Hieronymus Lommer, *Bergmännischer Beytrag zu der von der Königlichen Großbrittannischen Societät der Wissenschaften, auf das Jahr 1781 ausgestellten Preißfrage: Wie waren die Bergwerke bey den Alten eigentlich beschaffen und eingerichtet? Und läßt sich nicht nach angestellter Vergleichung derselben, mit den unsrigen, zum Vortheil des Bergbaues, und Hüttenwerke in unsern Zeiten, etwas von den Alten lernen?* (Freiberg, 1785).

[2] Blumenbach was an influential paleontologist before Cuvier, and professor of medicine at Göttingen University after 1778. Göttingen was one of the few universities that had incorporated scientific studies into the curriculum as part of a larger interest in cameralism. Donata Brianta, "Education and Training in the Mining Industry, 1750–1860: European Models and the Italian Case," *Annals of Science* 57 (2000), 275–278. See also André Wakefield, *The Disordered Police State: German Cameralism as Science and Practice* (Chicago: The University of Chicago Press, 2009).

sciences "in the time of [Georg] Agricola," which, Lommer argued, were suggestive but misguided inquiries grounded on traditions and superstitions rather than on reason. Miners had attributed healing and supernatural qualities to minerals, a "childish fairy tale" that was now replaced by "system and knowledge."[3] Mine surveying was "Devil's work" before joint-founder of the Academy Friedrich Wilhelm von Oppel's textbook on mine surveying (1749). Before proper concepts and labels were united with corresponding objects of nature, so-called mining language (*Bergsprache*) dominated in practice, the quaint and useful approach of the common man, which lacked the simplicity and clarity of enlightened concepts.[4] In the absence of proper reason and experiment, the miners imagined that the stars and planets had an effect on mineralogical phenomena, that mineral veins originated in the hot molten center of the earth, and that the dowsing rod provided reliable information. The rod was the epitome of all this nonsense, Lommer stressed, a powerful dream among "so-called knowers of mining" (*sogennante Bergverständige*) and even misguided scholars that lasted into the eighteenth century.[5] The miner of Lommer's day was embarrassed by continued use of the old dowsing rod and relied rather on "natural observations and sound reason."[6]

Lommer's statements indicate that original Academy professors distinguished themselves from earlier *Bergverständiger*, marginalized prospecting technologies such as dowsing, and identified with an international Republic of Letters and new ideal of *Wissenschaft*, as represented at Göttingen. Historians of earth science have echoed Lommer by stressing paradigmatic change at Freiberg. Rachel Lauden cites Freiberg professor Abraham Werner as the first to distinguish historical geology from mineralogy.[7] David Oldroyd invokes Thomas Kuhn's theory of paradigm change: "The Wernerian radiation gave the study of the earth its first paradigm, marking a separation from mineralogy. It was the means of escape from a pre-paradigm condition."[8] Martin

---

[3] Lommer, *Bergmännischer Beytrag*, 12.
[4] Lommer, *Bergmännischer Beytrag*, 15.
[5] Lommer, *Bergmännischer Beytrag*, 15.
[6] Lommer, *Bergmännischer Beytrag*, 36.
[7] Lauden, *From Mineralogy to Geology*, 104–5.
[8] Oldroyd, *Thinking about the Earth*, 103. See also William R. Albury and David R. Oldroyd, "From Renaissance mineral studies to historical geology, in the light of Michel Foucault's *The Order of Things*," *British Journal for the History of Science* 10 (1977).

Rudwick also notes the epochal influence of Werner in the emergence of geology as distinct from mineralogy or natural history.[9] Economic and mining historians have similarly presented Freiberg innovations as a watershed in mine engineering (see Introduction).

The new agenda for mining—cameralist, scientific, technological—was a powerful discourse that had significant impact. But we have discussed a different, longer-standing language in mining towns like Freiberg, and that was the language of the older *Bergverständiger.* Mining books, mining town sermons, mining reports, patronage documents, and even academic texts since the sixteenth century developed an image of the experienced miner who possessed a tacit knowledge of the mountain, and how to locate mineral ore. He read the signs of ore on the landscape, the effects of presumed mineral fumes on the vegetation and trees, and he used the dowsing rod. The famous Joachimistal pastor, Johann Mathesius, summarized these techniques in his 1562 collection of sermons: the independent prospector, he said, "needs…reason and the advice of loyal and experienced people, keeps his eyes out of his pocket…follows veins and rifts with the dowsing rod [*Ruthe*], minds the mineral vapors, pebbles, rocks, and trees, assays often to be sure, and digs and climbs down in the name and by the word of Jesus Christ."[10] Hans Uttman, a metal smith and mining official, added in his 1601 manual that mountain springs originated in mineral veins, and that expert diggers looked for a "sulfuric fatness" and a vitriolic or otherwise sharp taste in waters.[11] Mining masters and surveyors from Thölde to Beyer and Voigtel presented this knowledge in their mining books, while hired dowsers put it into practice.

The legacy of this *Bergwissenschaft* continued into Lommer's day and beyond, though he failed to acknowledge its contribution. This chapter argues that Freiberg became a site for the integration of new science and prospecting tradition, notwithstanding Enlightenment rhetoric to the contrary. Freiberg professors and mining officials after Lommer had little alternative to the prospecting of old, and they assimilated mining books filled with mining lore, and age-old digging and

[9] Martin J.S. Rudwick, "Minerals, strata and fossils," in N. Jardine, J.A. Secord, and E.C. Spary, eds., *Cultures of Natural History* (Cambridge, 1996). Rudwick provides a more nuanced account of Werner and his intellectual context in *Bursting the Limits of Time*.

[10] Mathesius, *Sarepta*, 847. On Mathesius, see Chapter Two.

[11] Uttman, *Bericht, von denen Ertz-Gebürgen*, DHSA, Loc. 36070, 2b.

dowsing techniques into the curriculum. In addition, they followed the latest scientific research into dowsing. Animal magnetism and galvanism suggested an enlightened science of dowsing, or rather, Romantic science based on *Naturphilosophie*. Freiberg came to embody an educational ideal that synthesized Enlightenment and Romantic currents. The famous Freiberg physicist, Ferdinand Reich, rejected Romantic science, but the Central Mining Office nonetheless asked him in the 1840s to examine a mine surveyor who claimed to experience 'galvanic excitement' with a dowsing rod. The dowser could appropriate a theory of galvanism that the physics community otherwise embraced.

We begin with the Mining Academy to see how the dowser became the antithesis of the Enlightened specialist, but how prospecting defied the Enlightenment agenda. Early Academy texts that otherwise trumpeted progress and new science also reveal that prospecting remained a stubborn problem. I then turn to the place dowsing found among physicists and chemists, in particular the work of the German physicist and *Naturphilosoph* at the Munich Academy of Sciences, Johann Ritter. When Professor Reich at Freiberg examined a local mine surveyor and dowser with an electrometer and galvanometer, he was testing the latest theory of Ritter's. The case brings us back to the field, where we witness a dowser combine electrical theory with his tacit knowledge and an older mining science.

The School on the Mountain

The Freiberg Mining Academy (f.1765), intended to train elite officials, was the culmination of a long-term effort to bring state management over Saxon mining. Freiberg scholars and officials had long imagined an institutionalized mining science to replace traditions and customs. In 1702, *Oberberghauptmann* Abraham von Schönberg secured funding from the Elector to instruct in mining-related inquiries and crafts. The resolution made a distinction between "mining sciences, smelting, and mine surveying," and digging expeditions (*Schürfen*), which it also funded.[12] Formal proposals for a school also dated to

[12] 'Resolution wegen Abstell-und Remedirung derer in Bergwercks-Sachen vorgekommenen und angemerckten Mängel und Gebrechen, sonderlich die Freybergische Revier betreffend, den 7. Januar Anno 1709,' in W.M. Schaffrath, ed., *Codex Saxonicus: chronologische Sammlung der gesammten praktisch-gültigen Königlich Sächsischen*

Schönberg in 1712, but the most developed came in 1746 from the chemist and chief superintendent of mines (*Bergkommisar*), Karl Friedrich Zimmerman. He described a break with the *Bergverständige* and his language:

> Since the time that one relied alone on digging and the dowsing rod, we have become rather removed from inspecting the land, mineral vapors, temperature of air, and position against the stars. The old knowers of the mountain [*Bergverständige*], those who first established the local regions, looked to these things and even named the strike of veins accordingly. We retain the expressions of these old fathers in the Mineral Kingdom, but we are ever more removed from an original understanding of the matter itself.[13]

Zimmerman cited the 1601 *Bericht* of the metallurgist Hans Uttman (printed in 1732), which was full of such wisdom. The new *Bergwissenschaft* would speak the language of new science, not Uttman's old "expressions."

The Seven Years War devastated Saxony and triggered a new wave of fiscal and administrative reforms that Leibniz might have admired. Privy counselor Thomas von Fritsch headed a reform committee (*Restaurationskommission*) charged with jumpstarting the Saxon economy, including agriculture, manufacture, and commerce. In 1763, the Elector appointed Friedrich Anton von Heynitz to reorganize mining. He represented two key factors behind the revival of Saxon mining at this time: the codification of the *Direktionsprinzip*, or centralized control by the state over the mining and metals industry, and the expansion and rationalization of the mining bureaucracy.[14] Heynitz soon became superintendent of mines, expecting from the mining office "almost martial subordination and complete confidence."[15] In this capacity, Heynitz secured state funding for the Academy, and

---

Gesetze von den ältesten Zeiten, vom Jahre 1255 an bis zum Schlusse des Jahre 1840 (Leipzig, 1842), 496.

[13] Christoph Friedrich Zimmerman, *Obersächsisches Berg-Academie, in welcher die Bergwercks-Wissenschaften nach ihrem Zusammenhang entworffen werden* (Leipzig, 1746), 102. A student of the Stahlian chemist at Freiberg, Friedrich Henckel, Zimmerman became *Bergkommisar* shortly before his death in 1747.

[14] Brianta, "Education and Training in the Miningn Industry," 272.

[15] Wakefield, *Disordered Police State*, 30. On Heynitz (1725–1802), see also Wolfhard Weber, *Innovationen im frühindustriellen deutschen Bergbau und Hüttenwesen Friedrich Anton von Heynitz* (Göttingen: Vandenhoeck & Ruprecht, 1976). On Fritsch and the broader economic reform in which Heynitz participated, see Horst Schlechte, *Die Staatsreform in Kursachsen, 1762–1763* (Berlin: Rütten and Loening, 1958).

headed a committee to undertake a thorough examination of all mining and smelting operations and administration. It issued a comprehensive 1767 report, citing numerous problems, from the lack of mining experience of Dresden bureaucrats, to widespread neglect, incompetence, and mismanagement in the field, to the inevitable effects of war. Like Zimmerman before him, Heynitz also mentioned the poor state of mining science, and he attacked dowsers in this regard. The presence of iron ore was enough indication of nearby silver- or copper-bearing deposits, maintained Heynitz, calling for the "banishment" of dowsing and a renewed commitment to "natural knowledge."[16] Mining was not a form of "gambling."[17] Dowsing had become a symbol of all that was wrong with the mining administration, and as Heynitz replaced a host of officials in the field with more enlightened Freiberg graduates, he surely intended men who did not promote such superstition.[18]

Reformers conceived the Academy as part of an economic agenda, but the curriculum also reflected a university reform spreading throughout Germany. Places such as Göttingen and Berlin introduced Baconian fields such as mining into an overwhelmingly theory-based curriculum, adapting these practical inquiries to the traditional goals of a university. The new *Wissenschaft* held that a scientific inquiry was one that a rational thinker could reduce to a small number of basic and logically ordered statements, which were more fundamental and universal than technical precepts or empirical observations. Based on Friedrich Gottlob Klopstock's *The German Republic of Scholars* (1774) and Immanuel Kant's epistemology, other scholars like Johann Gottlieb Fichte, Friedrich Daniel Ernst Schleiermacher, Friedrich Wilhelm Joseph Schelling, and especially Wilhelm von Humboldt added the Romantic belief that university inquiries should synthesize into a larger unity, and that the pursuit of knowledge was a process of self-discovery.[19] *Wissenschaft* was not merely technical or vocational

---

[16] "Diese ganz kurz gefaßten, mehr zuverläßigen Erfahrungen zeigen, theil die Nothwendigkeit immer fortzusenzender Untersuchungen der Landes-Gegenden... theils geben sie die Möglichkeit an Hand, wie man mit Verbannung der Wünschelruthe... sich weit edler und nüzlicher, mit Erweiterung der Natur-Kenntniße, beschäfftigen könne." In Baumgärtel, *Bergbau und Absolutismus*, 139.

[17] Baumgärtel, *Bergbau und Absolutismus*, 135.

[18] Wakefield, *Disordered Police State*, 37.

[19] Frederick Gregory, "Kant, Schelling, and the Administration of Science in the Romantic Era," *Osiris*, 5:2nd Series (1989).

knowledge, but also implied character or cultivation (*Bildung*). It was an indication of high status, and integrated the educated upper middle-class with the nobility.[20] This university ideal influenced scientific academies, where there was a growing rift between learned and craft knowledge that would separate further into professional science societies and technical schools, respectively, through the nineteenth century. As an example, the Bavarian Academy of Sciences at Munich was hostile to the optics technician and prospective member, Joseph von Fraunhofer, because he lacked university training.[21]

The Freiberg Mining Academy had originally challenged contemporary universities by providing prospective civil servants with a more technical-oriented education and by subordinating professors to the state, but Freiberg also advanced the broader knowledge reform. Historians of mining have yet to analyze the transformation of the Academy from a small technical school to more 'Humboldtian' center of study, when it attracted Hardenberg (Novalis), Humboldt, and Goethe. The founding of a lower-level mining school (*Bergschule*) at Freiberg in 1777 for diggers, assayers, and other 'hands-on' jobs reflected the distinction that universities such as Göttingen advanced between new *Wissenschaft* and the crafts. The Academy drew more from Kant than Schelling, but Romantic authors would praise Abraham Werner as a *wissenschaftlicher Künstler*, or academic artist;[22] and Halle and Berlin, where the Romantic current was strong, had considerable influence at Freiberg. The curriculum provided practical training and required students to tour the mines, but it also encouraged officials-in-training to co-opt useful artisanal knowledge and construct a *Bergwissenschaft* informed principally by new ideals. Training elite mining officials here would involve both technical skill and

[20] Thomas Broman, "University Reform in Medical Thought at the End of the Eighteenth Century," *Osiris* 5:2nd Series (1989). Charles E. McClelland, *The German Experience of Professionalization: Modern Learned Professions and their Organizations from the Early Nineteenth Century to the Hitler Era* (New York, 1991).

[21] Myles W. Jackson, *Spectrum of Belief: Joseph von Fraunhofer and the Craft of Precision Optics* (Cambridge, MA, 2000).

[22] In *Die Lehrlinge zu Sais*, Novalis, who studied at the Freiberg Academy, likely intended the Temple at Sais to be a metaphor for the school, and the Temple's master to be Abraham Werner. The old miner in *Heinrich von Ofterdingen* is probably Werner again. Noah Heringman, "The Rock Record and Romantic Narratives of the Earth," in *Romantic Science: The Literary Forms of Natural History*, Noah Heringman, ed. (Albany, NY: The State University of New York Press, 2003), 68.

character: "The academic politics of both the German Enlightenment and Romanticism...ultimately concerned training civil servants."[23] Both strains are visible in the Academy's history.

Heynitz originally expected Freiberg professors to embody a more decidedly Enlightenment ideal that pit miner tradition against progress. The first professor of metallurgical chemistry, Johann Friedrich Wilhelm von Charpentier, expressed frustration with workers in a 1799 piece on *Geognosie*. He was proud of the mining tradition at Freiberg, but confident in a more scientific approach to metallurgical phenomena. Suggesting that earth science could satisfy an ideal of mathematical clarity, Charpentier began with praise for Galileo, Kepler, and Newton.[24] The "common miner" then emerged as a source of ambiguity for Charpentier. Insofar as the description of mineral veins remained in common mining language, the subject was shrouded in darkness.[25] The majority of miners did not distinguish mineral veins from surrounding earths well enough. A proper account required a more sophisticated understanding of the ore.[26] Charpentier repeatedly contrasted the clearer and more distinct language of natural science with the prejudiced views and expressions of the workers: "I have described what I have seen, and nothing else...and just as it would appear to anybody with no preconceived notion."[27]

The title-emblem of Charpentier's 1778 *Mineralogical Geography of the Saxon Lands* is significant here for depicting the dowsing rod centrally [Fig. 8]. A muse holds an official letter. To the right, the bust of the Elector sits atop a Roman column. To the left, blindfolded Prejudice stumbles, as an angel breaks his dowsing rod. Charpentier's message is clear: Enlightened cameral management over mineral resources overcomes blind superstition in Saxon mining. Heynitz's reform committee had given Charpentier the assignment to create a mineralogical map of Saxony. The emblem echoed Heynitz's call to banish dowsing and ground prospecting on a more systematic understanding of visible geological characteristics. Heynitz had proposed iron-bearing earth to be the telltale sign of nobler ore, and Charpentier was equally

[23] William Clark, *Academic Charisma and the Origins of the Research University* (Chicago: University of Chicago Press, 2006), 446.

[24] Johann Friedrich Wilhelm von Charpentier, *Beobachtungen über die Lagerstätte der Erze, Haupsächlich aus den Sächsischen Gebirgen* (Leipzig, 1799), Preface.

[25] Charpentier, *Beobachtungen*, 38.

[26] Charpentier, *Beobachtungen*, 49.

[27] Charpentier, *Beobachtungen*, Preface.

confident that better knowledge of the earth would prevent unnecessary digging and expense.[28]

Fig. 8. An angel breaks the dowsing rod in the title image to Johann Friedrich Wilhelm von Charpentier's *Mineralogische Geographie der Chursächsischen Lande* (1778), showing the triumph of reason over superstition.

[28] "Die genauere Kenntnis der Steinarten, woraus die Gebürge bestehen, das Verhalten der Gänge und anderer Lagerstätte der Erze, ja die Beschaffenheit der Erze

The most authoritative and renowned voice at Freiberg was Abraham Werner (1750–1817). Enlightenment scholars and *Naturphilosophen* alike saw *Wissenschaft* in Werner's works and lectures. His classification scheme and advancement of a neptunist theory—that rocks lay in expansive and distinct formations that a process of sedimentation had laid in succession—generated a flurry of field research in Saxony and abroad.[29] Not surprisingly, such a progressive thinker could be dismissive of mining traditions. In his 1791 *New Theory on the Origin of Mineral Veins*, Werner listed all major contributions to the subject of mineral generation—from Pliny, to Agricola, to Johann Becher, to Charpentier—which Werner categorized into four: theories that mineral veins originated at Creation, theories that they were branches of an enormous mineral tree, theories that they were hollow passages that filled with mineral substance, and theories that they developed from the earthly material that surrounded them. Werner settled on a synthesis, believing that hollow passages filled with contiguous materials that chemically modified over time. He explicitly excluded the work of Hans Uttman (1601), mine manager Georg Engelhard von Löhneiß (1617), and other miner books after Agricola from the discussion: "What they offer is too unimportant to deserve mention." Werner meant in particular that they gave too much causal significance to the stars.[30] He denied that astrology and dowsing still had currency at his contemporary Freiberg: "The entire old belief of the influence of the sun and planets on mineral veins belongs, together with the now rather forgotten dowsing rod [*Wünschelruthe*], to the astrological mysticism of old times...and hardly deserves mention, the name of theory, and certainly not refutation; because the claims are totally inconsequential and incoherent, and this superstitious belief is long forgotten."[31]

---

selbst, macht uns wenigstens behutsam, und lehrt uns, nicht überall und in jeder Steinart Erz zu suchen und sich einzubilden, da viel zu finden, wo nichts oder wenig zu finden ist." In Guntau, *Die Genesis der Geologie*, 48.

[29] On Werner, see André Wakefield, "Abraham Gottlob Werner and the Cameralist Tradition in Freiberg," *Freiberger Forschungshefte*, D207 (September, 2002); Walther Herrmann, "Die Zeit Abraham Werners in Freiberg," *Freiberger Forschungshefte* D2 (Berlin, 1953); and Hans Jürgen Rösler, ed., "Abraham Gottlob Werner: Gedenkschrift aus Anlaß der Wiederkehr seines Todestages nach 150 Jahren am 30. Juni 1967," *Freiberger Forschungshefte* C223 (Leipzig, 1967).

[30] "Sie sezzen insgesamt den Hauptgrund der Veredlung der Gänge, theils auf die Lage des Gebirges gegen die Sonne, theils auf den Einflus der Gestirne." Abraham Werner, *Neue Theorie von der Entstehung der Gänge, mit Anwendung auf den Bergbau, besonders den freibergischen* (Freiberg, 1791), 16.

[31] Werner, *Neue Theorie*, 170.

Charpentier and Werner's statements, like Lommer and Heynitz's, were in accord with other scholarly remarks on miners across Europe. More local pamphlets and calendars spread Enlightenment thinking through mining towns. These included the *Harzmagazin* in the Harz Mountains, and the Marienberg *Bergwerkscalendar* in the Ore Mountains. A short 1775 essay in this calendar entitled, "On the Superstitions of the Miner," was a summary of major beliefs and practices that fell under this category, including astrology, mining spirits, treasure hunting manuals, and the dowsing rod.[32]

But did institutionalized *Wissenschaft* really displace the *Bergwissenschaft* of old, as Heynitz and Werner claimed? That another issue of the same Marienberg calendar included Agricola's entire chapter on digging and dowsing suggests that the scholars of mining did not possess an alternative prospecting practice, and that learned argument against the dowsing rod had not advanced beyond Agricola's ambiguous statements.[33] The prospecting and digging (*Schürfen*) that Agricola discussed involved the tracking and analysis of mountain waters and springs, the study of alluvial mineral fragments (*Geschiebe*), and the determination of the supposed effects of mineral vapors and heats on vegetation aboveground. Agricola rejected dowsing that involved magical incantations or characters, but he issued no definitive statement against the rod per se, describing and depicting legitimate dowsing in some detail. The fields taught at the Academy included pure and applied mathematics, land and mine surveying, mine engineering, natural history of minerals, chemical mineralogy, and smelting.[34] Some of these subjects were new or highly progressive, such as chemical mineralogy and mine surveying; but digging, the search for the 'natural signs' of mineral ore and its discovery, remained steeped in mining lore and tradition. *Schürfen* would find a place within the new Academy curriculum, and that meant continued toleration or reevaluation of the rod and its bearers.

The Mining Academy incorporated *Schürfen* into mine engineering (*Bergbaukunde*) more broadly, a field that also included the study of

[32] "Würde wohl die Wünschelruthe...die Bergleute so lange getäuschet haben, wenn sie hierbey nicht so leichtgläubig gewesen wären?" Gotha Forschungbibliothek (GFB), Math 8° 1260/2, year 1775.

[33] GFB, 8° 1260/2, year 1778.

[34] Walter Hoffman, *Bergakademie Freiberg* (Frankfurt am Main: Verlag Wolfgang Weidlich, 1959), 43.

the earth (*Lagerstättekunde*), boring, and mine construction and mechanics (hydraulics and ventilation).[35] Werner and others lectured on *Bergbaukunde*, though it never received the same treatment as more progressive fields had.[36] The textbook from which Werner lectured exposed this weakness in the curriculum. Johann Gottlieb Kern's *Bericht vom Bergbau* (1772) was intended to replace prior mining books, as joint-founder of the Academy, Oppel, explained in the preface. He cited three authorities of old: Georg Agricola's *De re metallica* (1556), Georg Löhneiß's *Bericht vom Bergkwerck* (1617), and Balthasar Rößler's *Bergbauspiegel* (1700). All three works included substantial sections on *Schürfen* and dowsing. Oppel praised their efforts but considered them merely stepping-stones toward a comprehensive teaching.[37] However, when Kern turned to *Schürfen*, his text lost its enlightened luster and resorted to the tacit knowledge of old. In keeping with Academy rhetoric, Kern excluded the dowsing rod, but what remained was little different from that explained in Agricola and repeated in Uttman, Löhneiß, and Rößler. *Schürfen* was, said Kern, the study of alluvial mineral fragments embedded in topsoils, and the inspection of earths and grasses, particular herbs, and warm springs, which one examined for a salty or vitriolic taste, and an oily fattiness. He also referenced the age-old belief that patches of evaporated dew or melted snow revealed the presence of mineral vapors, and the minerals from which they came, though he considered that an uncertain sign.[38] This knowledge was the product of centuries of experience locating mineral ore.

Other Academy texts acknowledged their debt to an older *Bergwissenschaft* more openly. The chief overseer of Saxon mining after 1801, and first Academy graduate, Friedrich Wilhelm Heinrich von Trebra

[35] Lommer taught the subject until 1771, when Werner took over. He defined the field as such: "Lehre vom Gebrauch des Grubenkompasses, Bestimmung des Streichens und Fallens des mineralogischen Lagerstätte, die Lehre von der Beurteilung der Bauwürdigkeit ganzer Reviere und ganzer Gebirge, die Lehre von der Aufsuchung sowohl unerschrotener und noch unbekannter, als auch bekannter, desgleichen auch verlorner metallischer Lagerstätten, wohin ich zugleich die Aufsuchung und Beurteilung der Geschiebe, des Schürfen und Übberröschen und den Gebrauch des Erdbohrers erklärt habe, ferner die Lehre von der Veranstaltung eines neu angehenden Grubenbaues." In Joachim Wrana, ed., *Bergakademie Freiberg: Festschrift zu ihrer Zweihundertjahrfeier am 13. November 1965* vol. 1 (Leipzig, 1965), 118.

[36] Wrana, *Bergakademie Freiberg*, 120.

[37] Johann Gottlieb Kern, *Bericht vom Bergbau* (Leipzig, 1772), Preface.

[38] Kern, *Bericht*, 29.

(1740–1819), advanced the Freiberg agenda, but he admitted the limits of new science and reason in the case of prospecting. Trebra, hand-picked by Heynitz to become the new *Bergmeister* at Marienberg, adopted the official rhetoric, which included praising the introduction of higher learning into what had otherwise been merely a craft:

> Mining had never been acknowledged as a scientific [*Wissenschaftlichen*] enterprise, but rather disregarded as a craft, or at best an art, to which… other than miners, only a few scientific men wished to contribute. The Mining Academy…remedied this evil."[39]

However, we know that Trebra used the mining sermons of Johann Mathesius (1565), the chronicles of Saxon mining of Peter Albinus (1590), and the mining books of Balthasar Rößler (1700) and Christoph Herttwig (1710) in the lecture hall, all of which described miner lore and digging knowledge, and included even detailed explanations of dowsing.[40]

Trebra's first major publication, *Experience of the Interior of Mountains, as Gathered through Observations*, treated the problem of *Schürfen* head-on, where Trebra expressed an amount of frustration with learned works on the matter. He claimed that to present the subject adequately, he had to employ miner language (*Bergsprache*), and would therefore have to explain this terminology for the reader.[41] Rather than ridicule this language or dialect, as Lommer had, Trebra acknowledged its importance for his learned audience. There would be no discussion of the origin of veins, or similar theoretical issues in his text, moreover, since the amount of experience gathered thus far did not warrant it, nor would that serve the immediate needs of miners.[42] Trebra was equally critical of recent trends in mineralogy that lost sight of usefulness in mining. He censured the famous Swedish chemist and mineralogist, and discoverer of nickel, Axel Friedrich Cronstedt. In his "Essay toward a new Mineralogy," translated and annotated by Werner in 1770, Cronstedt, the "father of this science,"

---

[39] In Baumgärtel, *Bergbau und Absolutismus*, 98.

[40] Walther Herrmann, *Goethe und Trebra: Freundschaft und Austausch zwischen Weimar und Freiberg* (Berlin, 1955), 15.

[41] "Ich habe mich oft der Sprache des Bergmanns bedienen müssen…diese und der Zusammenhang des Ganzen, sollen hoffentlich dem aufmerksamen Leser nirgends Unverständlichkeit übrig lassen, wenn er auch gleich kein Bergmann ist." Friedrich Wilhelm Heinrich von Trebra, *Erfahrung vom Innern der Gebirge, nach Beobachtungen gesammlet* (Leipzig, 1785), Preface.

[42] Trebra, *Erfahrung*, Preface.

exposed a lack of experience on the mountain and appealed more to collectors and learned men than to miners. Practical men needed rules by which they could find mineral ore, maintained Trebra, and Cronstedt (and Werner) could only concede how inexact prospecting was.[43]

Trebra wished prospecting to become a science grounded in miner experience. On the one hand, seeing that *Schürfen* still passed orally "from father to son," he called for more order and the replacement of popular wisdom with reliable or scientific knowledge.[44] Trebra's basic prospecting rule was that mineral veins shared their origins with water sources, and that miners might therefore look for wet crevices and other damp regions.[45] He accredited his renowned success at locating silver veins at Marienberg to following that simple rule.[46] But on the other hand, Trebra appealed to classic miner beliefs on natural signs as presented in Mathesius or Rößler: fallen trees that revealed mineral ore, alluvial mineral fragments that could lead to veins, and particular earths known to lay contiguous with ore.

A dual commitment to Academy knowledge and prospecting experience explains Trebra's ambiguous statements on the dowsing rod. He echoed fellow Academy scholars like Charpentier and Werner by presenting the practice as the epitome of popular superstition in mining. Trebra included a map of veins as charted by a dowser in 1709, and the absurdity of the practice to Trebra was clear by the high number of charted veins.[47] He also criticized learned men who entertained dowsing merely out of desperation with a failing venture.[48] Yet, Trebra's explanation of dowsing was surprisingly detailed given his negative comments, and he clearly drew from Rößler. In Trebra's diary for 1767–1779, when he was *Bergmeister* at Marienberg, moreover, we see that he was more open-minded than his later *Experience* would

[43] Er [Cronstedt] befriedigte vollständiger den blossen Kenner, den Sammler, über die Kennzeichen, über die Eigenschaften der mineralischen Körper; aber den arbeitenden Bergmann, der diese Mineralien aussuchen soll, liess er über die Oerter, wo die Natur sie hingelegt hat, doch noch ohne ausgearbeitete Hülfsmittel gegen die Erwartung des blossen Ohngefährs." Trebra, *Erfahrung*, 4.

[44] "Es ging hier, wie es mit blossen Sagen allemal gehet, sie pflanzen sich wohl fort, vererben sich von Vater auf Sohn, werden aber selten ganz gründlich durchgedacht, erprüft, und berichtiget." Trebra, *Erfahrung*, 5.

[45] Trebra, *Erfahrung*, 9.

[46] Trebra, *Erfahrung*, 188.

[47] "Wo haben wir jemals ein Gebirge so besäet mit Gängen gefunden? Ich sage kein Wort mehr hierüber, das frappant Unwahre fällt allzusehr sogleich in die Augen." Trebra, *Erfahrung*, 67.

[48] Trebra, *Erfahrung*, 4.

suggest. For example, he once fined a foreman and "old magician" one week's wage for having failed to dowse a region of promising earth. It seemed that the magical rod had suggested the effort would be in vain. Duly censured, the man continued his work before Trebra, until they discovered traces of galena (*Bleiglanz*) and ultimately silver. Now the rod dipped strongly, Trebra sarcastically noted. More concerned with the man's negligence than with his dowsing per se, Trebra had considered the man stubborn or difficult, but now he was a "good old man."[49] On another occasion, an otherwise "trustworthy and hard-working" digger was caught with a piece of stolen mineral earth. His only defense was that he intended merely to use the piece in dowsing. He had actually been witnessed holding the ore beside an extended rod, attempting to increase its affinity for the earth below. The digger was fired. That Trebra spoke highly of the man and stressed the small size of the stolen piece, intended only for dowsing, suggested that Trebra sympathized with the poor digger.[50] Trebra also exposed his familiarity with the practice here by referring to the 'rod' (*Rute*), like practitioners and officials over the centuries had, rather than to the 'divining-' or 'wishing rod' (*Wünschelrute*), as had Heynitz and Werner.

An appendage to the *Diary* devoted to *Schürfen* helps explain why the skeptic Trebra tolerated dowsing in the late 1770s. He recognized that the Academy could not claim that prospecting was a science. Mining was otherwise a *Wissenschaft*, Trebra began, which drew on Christian Gellert's metallurgical chemistry and Johann Suckow's and Johann Krüger's physics (*Naturlehre*), among other new sciences. But digging was an accumulation and combination of tacit knowledge and scholarly insights, collective wisdom that Trebra classified into five branches: study of the general landscape and rock formations, examination of local vegetation, survey of the layout of mountains, reading

[49] "Ich fand weiter hin nie solche Widerspänstigkeit mehr, bey diesem übrigens guten alten Manne, die ich doch wohl auch dießmal, mehr der bisher gewohnten allgemeinen Verwilderung, als seiner Ruthengeherey zuschreiben konnte." Friedrich Wilhelm Heinrich von Trebra, *Bergmeister-Leben und Wirken in Marienberg, vom 1. Decbr. 1767. bis Augusti 1779* (Leipzig, 1990), 183–5. Originally published at Freiberg, 1818. Also, August Friedrich Wappler, "Oberberghauptmann von Trebra und die drei ersten sächsischen Kunstmeister, Mende, Baldauf, und Brendel," *Mitteilungen des Freiberger Altertumsvereins* 41 (1905): 76.

[50] "Er kam sogleich in Untersuchung, und ob gleich das Corpus delicti sehr klein war, und er vollen Beweis dafür beybrachte, daß er das Stüffchen nur weggenommen habe, um es beym Ruthengehen zu gebrauchen, denn dabey hatte man ihn mit dem Stüffchen in der Hand neben der Ruthe betroffen, so half ihm dieses doch nicht." Trebra, *Bergmeister-Leben*, 356.

of old and new authors, and consultation of "still living experts" in the field.[51] More so than Werner, Trebra acknowledged the gap between new theory and practice at the Academy, particularly in the case of prospecting.

Academy scholars did not treat *Schürfen* in general and dowsing in particular prominently or even sympathetically. They certainly expected the light of science and progress to shine through the earth's strata and render the knowledge of dowsers obsolete: "Maybe I will find some enlightenment [*Aufklärung*] in the fogs that continue to surround this science with a curtain impenetrable to many eyes."[52] In the meantime, they continued consulting the books of old and others more expert on the mountain than themselves, and that might include dowsers.

The Electric Rod

Rachel Lauden rightly contends that geological knowledge at the new Freiberg Academy had little bearing on mining practice. She cites prospecting in particular as an arena still based on "time-honored methods," including dowsing.[53] Hugh Torrens similarly argues that 'scientific prospecting' in coal mining based on the stratigraphic column was a British accomplishment of around 1805, but remained a gentlemanly and scholarly pursuit: "The practical men did not want to know."[54] Drilling technology was available, but expensive, and it did not have marked influence in mining until well into the nineteenth century.[55] More promising in practice were the new sciences of electricity and galvanism. These *did* suggest a science of prospecting, insofar as a number of prominent French, Italian, and German physicists began experimenting with pendulums and dowsing rods to detect the electrical or electromagnetic potential of mineral earth. An electrical theory

---

[51] Trebra, *Bergmeister-Leben*, 585.

[52] Trebra, *Bergmeister-Leben*, 586.

[53] "Prospectors relied on the time-honored methods of following surface deposits underground, examining hillsides after rain to see if unusual deposits had been exposed, tasting spring water for unusual metallic tastes, running trial sinkings and borings, and (most dubiously) using the divining rod." Lauden, *From Mineralogy to Geology*, 54.

[54] Hugh Torrens, "Some Thoughts on the Complex and Forgotten History of Mineral Exploration," *Journal of the Open University Geological Society*17 (1997): 20.

[55] Multhauf, *Neptune's Gift*, Chapter Eight.

of dowsing synthesized new science and *Schürfen* in a way Lommer never recognized.

Germany inherited the new dowsing theory from France and Italy. It originated with the French physician and Royal Inspector of Mineral Waters, Pierre Thouvenel (1747–1815). When Anton Mesmer commanded attention by 'magnetizing' water baths in Paris, Thouvenel discovered mineral waters near Contrexéville with the help of a peasant dowser named Barthelemy Bléton.[56] Thouvenel published his findings in a 1781 text, in which he replaced earlier theories of dowsing (mineral vapors, occult qualities, corpuscularism) with a theory of 'subterranean electricity' (*électrométrie souterraine*), an expression of the more fundamental animal magnetism Mesmer discussed.[57] Bléton's fits above mineral and water sources were, argued Thouvenel, qualitatively similar to the medical crises Mesmer witnessed in his magnified baths, and the law-like movements of the dowsing rod—spiraling clockwise or counterclockwise—suggested the polar action (positive or negative charge) of electricity. A commission was appointed in 1782 to investigate these claims, which included such famous students of electricity and Enlightenment thought as Benjamin Franklin, chemist Claude von Berthollet, the Baron d'Holbach, and physician Joseph-Ignace Guillotin. A flurry of opinions followed, and by the time Thouvenel and Bléton migrated to Italy, the dowser had stood before Condorcet, Bossuet, and Diderot.

In Italy, Thouvenel discovered yet another French dowser, Joseph Pennet, and found an audience with the famous biologist, Lazzaro Spallanzani. Inspired finally by Luigi Galvani's 1786 demonstrations with frog legs of an animal electricity (galvanism), Thouvenel associated this biological or chemical sort with his own proposed subterranean electricity. According to him, animal electricity (demonstrated with frogs) and mineral electricity (demonstrated with dowsers) seemed to "explain [one another] and to confirm [one another] mutually."[58] The contraction of frogs' legs paralleled the tremors and

[56] Lynn, "Divining the Enlightenment." On popular interest in electricity during the Enlightenment, see Oliver Hochadel, *Öffentliche Wissenschaft: Elektrizität in der deutschen Aufklärung* (Göttingen, 2003). Also Robert Darnton, *Mesmerism and the End of the Enlightenment in France* (Cambridge, MA, 1968).

[57] Pierre Thouvenel, *Mémoire Physique et Médicinal montrant des Rapports Evidents entre les Phénomènes de la Baguette Divinatoire, du Magnétisme Animale et de l'Electricité* (Paris, 1781).

[58] Quoted in Walter Bernardi, "The Controversy on Animal Electricity in Eighteenth-Century Italy: Galvani, Volta, and Others," *Nova Voltiana: Studies on Volta*

convulsions experienced by the dowser. Echoing Galvani, Thouvenel began calling his dowsers 'electrometers.'[59]

Thouvenel's claim that individuals of sensitive perceptual constitution functioned as electrometers circulated through learned journals of the day, including the German. As early as 1783, the *Hannoverisches Magazin* detailed Thouvenel's work with Bléton, and in 1791, the *Annalen der Geographie und Statistik* translated a letter of Spallanzani's on Thouvenel's more developed theories. Spallanzani proved himself an intimate of Thouvenel and Pennet's, and a close but skeptical observer to the end. The *Neue allgemeine deutsche Bibliothek* and the *Magazin für das Neueste aus der Physik und Naturgeschichte* then published reviews of the first German translation of Thouvenel in 1794 and 1795, respectively.[60]

The physicist, Johann Wilhelm Ritter (1776–1810), developed Thouvenel's theory. With Ritter, we transition back to the intellectual contexts of Weimar, Jena, Halle, Leipzig, and Freiberg. Ritter is best known in the history of physics as the discoverer of ultraviolet light (1801), for improving on Volta's pile of copper/zinc plates (or battery), and for using it to measure chemical change, or electrolysis. Strongly influenced by Friedrich Wilhelm Joseph von Schelling's philosophy of nature (*Naturphilosophie*) and the Romantic critique of Newtonian mechanics, moreover, Ritter also believed that physics engaged the spirit, reconciled man with nature, and provided an understanding of unity and harmony in nature. In his effort to prove that a galvanic force pervaded all organic and inorganic matter, Ritter invoked such themes from Schelling's philosophy as dualism, polarity, periodicity,

and his Times I (2000): 113. See also Lucia De Frenza, *I sonnambuli delle miniere: Amoretti, Fortis, Spallanzani e il dibattito sull'elettrometria organica e minerale in Italia (1790–1816)* (Florence, 2005).

[59] The electrometer was an instrument to measure static electricity developed by William Henley in 1770.

[60] Adolf Moller, "Nachricht von Bleton, einem Jüngling von erstaunlich empfindlichen Nerven, vermöge welcher, wie auch durch den Gebrauch der sogenannten Wünschelruthe, er unterirrdisches Wasser entdecken und anzeigen kan," in *Hannoverisches Magazin* 21 (1783), 945–960. Lazzaro Spallanzani, "Ueber das Vermögen, unterirdische Gewässer, verborgene Metallmassen und Mineralen zu spüren und sie zu verfolgen," in *Annalen der Geographie und Statistik* 2 (1791), 554–567. "Ueber unterirdische Elektrometrie, nebst einigen sie betreffenden, in Italien und in den Alpen vorgenommenen Versuchen. Aus dem Franz. übersetzt," in *Neue allgemeine deutsche Bibliothek* 10:2 (1794), 511–13. "Nachricht von der unterirdischen elektrometrie des Herrn Thouvenals," in *Magazin für das Neueste aus der Physik und Naturgeschichte* 10:1 (1795), 144–159.

and a vital and developmental spirit in nature, and the microcosm/macrocosm distinction.[61]

A lifelong interest in *Naturphilosophie* increased in Ritter's later years. In 1806, Ritter was correlating the rhythms of the voltaic pile with astronomical events. That was the year that Ritter, newly installed at the Bavarian Academy, received word from a fellow member, the mineralogist and crystallographer (and former Werner student), Christian Samuel Weiss, that a twenty-one year old farmer had successfully located hidden coins with a dowsing rod, and that he had learned his art from none other than Thouvenel and Pennet. Still looking for new ways to demonstrate the qualitative similarity between physical and biological electricity, and the unity of man with nature, Ritter asked the Academy for a travel stipend. So began two years of intensive experimentation with Campetti.

Just like Thouvenel before him, Ritter consulted both Carlo Amoretti and Volta, among other learned men in Italy. Ritter met Volta in 1807. The men discussed electricity and Ritter's variations on the copper/zinc pile Volta had invented. To all accounts, the meeting was cordial, though Ritter lamented that he could not interest his Italian colleague in his latest research: "He is pleased to hear about anything that one points out to him, and always with childlike fervor. However, one may not ask him about those issues he too often sets aside. So, for example, one could not talk to him about Campetti, etc."[62] Volta too commented on this meeting with the curious German *Naturphilosoph*: "In one word his views are too transcendent, he runs and pushes things too far."[63]

Ritter brought his dowser back to Munich for more focused experimentation, beginning with pendulum experiments. The instrument consisted of a rock of pyrite (*Schwefelkies*) on a string of human hair or raw silk, hung over an object resting below, generally magnetite

---

[61] A compact introduction to Schelling's philosophy and its influence in the sciences is Nicholas Jardine, "*Naturphilosophie* and the kingdoms of nature," in N. Jardine, J.A. Secord, E.C. Spary, eds., *Cultures of Natural History*. See also Reill, *Vitalizing Nature*, chapter 5, Olaf Breidbach and Paul Ziche, eds., *Naturwissenchaften um 1800: Wissenschaftskultur in Jena-Weimar* (Weimar, 2001), and Robert J. Richards, *The Romantic Conception of Life: Science and Philosophy in the Age of Goethe* (Chicago, 2002). On Ritter's physics, see Walter D. Wetzels, "Johann Wilhelm Ritter: Romantic physics in Germany," in Andrew Cunningham and Nicholas Jardine, eds., *Romanticism and the Sciences* (New York, 1990).

[62] Richter, *Das Leben*, 140.

[63] Kleinart, "Volta and the German Controversy," 37.

(lodestone) or another mineral or metal. The 'Water-Feeler' (*Wasserfühler*), 'Metal-Feeler' (*Metalfühler*), or 'Electrometer,' as Ritter called his subjects, demonstrated receptivity when the pendulum oscillated. According to him, the subject conducted a force between the hanging and placed objects that was otherwise neutralized in nature.

Ritter carried out hundreds of such experiments, believing to observe oscillations of law-like regularity, as Thouvenel had. Above magnetite the hanging pyrite circulated either clockwise or counterclockwise depending on which pole of the magnetite it neared. Ritter lay other minerals below the hanging pyrite. They too seemed to show magnetic or electrical poles, as the pyrite in Campetti's hand oscillated in different ways above each pole. Copper, silver, zinc, and even water exposed their polar natures through the pendulum experiments. Ritter did not stop with the mineral kingdom. He began experimenting with the polar nature of various organic and inorganic substances, including apples, oranges, eggs, and even the human body. Ritter claimed to demonstrate the polar properties of various body parts: the human head had the same polar properties as zinc, whereas the soles of the feet responded like copper. The pendulum oscillated above all fingertips.

Ritter devised a second group of experiments using a device he called a *Balancier*. The standard version was a copper bar some six inches long and half an inch thick, though its composition, shape, and size could vary. The experiment consisted of a presumed conductor standing over a particular mineral or metal while balancing the bar on the fingertips. The bar tipped one way or another with varying intensity, or it remained still, depending on which substance lay below the subject and any number of other variables. These included the polar nature of the chosen finger, the composition and shape of the bar and substance below, the subject's respiration and other physiological activities, the position of the arm and legs, and the subject's sex—different sexes manifested opposing motions.[64] In all this, Ritter was interested in the polar nature of the supposed force coursing through his subjects, as demonstrated by the supposed law-like motions of the Balancier.

Ritter concluded that he was working with a natural force coursing through Campetti's body, qualitatively similar to galvanism. He called

[64] Gehlen, "Nachrichten," 116–7.

his force "siderism" (*sidus*=star), maintaining that it resembled the gravity of planetary bodies. The movements of the dowsing rod, pendulum, or metal rods were equivalent at a microcosmic level of what occurred in the universe at a macrocosmic level.[65] Ritter described his interest in Thouvenel's work, his entire experience with Campetti, and the theory of siderism in a work of 1808.[66]

Did Freiberg scholars know about Ritter's research? While Ritter never visited the Mining Academy, nor corresponded with Werner, Trebra, or Reich, Ritter's influence clearly extended to the very doors of the Academy. None of the Freiberg professors were themselves *Naturphilosophen*, but a number of Werner's famous students were: the Romantic author, Friedrich von Hardenberg (Novalis), left Jena in 1797 to study under Werner at the advice of Ritter.[67] The physician, Gotthilf Heinrich von Schubert, also came to Freiberg after exposure to Schelling and Ritter at Jena. Schubert referenced Ritter's work in a text on geognosy and mining: discussing *Schürfen*, the author mentioned the 'old experiences' concerning melted snows, evaporated dews, discolored and disfigured vegetation and trees, as well as the latest research on "metal-feelers," which he found very promising.[68]

---

[65] "Siderismus bleibt sein Name, denn wirklich ist es nur der Sternenlauf, der sich in allen jenen sonderbaren Bewegungen von Baguetta, Pendel, Degen, Früchten, Magneten, Metallen, usw. wiederholt, indem jeder Körper der anorganischen Natur, in Conflict mit dem vollendetsten Microcosmus, dem Menschen, selbst zu einer, um was immer für ein Centrum als Sonne, laufenden und sich drehenden Erde zu werden sucht." Ritter to Karl von Hardenberg (the author Novalis's brother), in Armin Hermann, *Die Begründung der Elektrochemie und Entdeckung der Ultravioletten Strahlen von Johann Wilhelm Ritter* (Frankfurt am Main, 1968), 18.

[66] Johann Wilhelm Ritter, ed., *Der Siderismus* I (Tübingen, 1808). Positive and negative reviews of these experiments appeared in contemporary journals, among them *Journal für die Chemie und Physik* [JCP] published by Ritter's friend and colleague at Munich, Adolph Ferdinand Gehlen. See "Nachrichten von den neuern durch Francesco Campetti wieder rege gewordenen Versuchen über Pendel, Baguette, u.," *JCP* 4:1 (1810).

[67] "Der Jenaer Physiker Ritter wies ihn auf Werner hin." Bernd Klengel, "Ultaviolette Strahlen und Ladungssäule: Zur Rezeption von Entdeckungen Johann Wilhelm Ritters in Frankreich," in Breidbach and Ziche, *Naturwissenschaften*, 52. See also Alexander M. Ospovat, "Romanticism and German Geology: Five Students of Abraham Gottlob Werner," *Eighteenth Century Life* 7 (1981–2): 105–117, and Nicholas A. Rupke, "Caves, Fossils and the history of the Earth," in Cunningham and Jardine, eds., *Romanticism and the Sciences* (New York: Cambridge University Press, 1990).

[68] "Was jene Arten der Aufsuchung metallischer Gänge und Lager betrifft, wobei die Nähe der Metalle mittelst eines krankhaft gereizten Nervensystems empfunden, oder durch das leise Zucken der gewaltsamen gespannten Fingermuskeln, bei einer natürlichen Reizbarkeit für Metalle merklich wird, mit andern Worten das sogenannte Metallfühlen und Wünschelruthen-Schlagen, so dürfen die Erfahrungen, auf welche

Geologist Henrik Steffens studied under Werner from 1799 to 1802, and synthesized his geognosy with Schelling's thought. Steffens was a more outspoken (if also critical) admirer of Ritter.[69] Alexander von Humboldt studied under Werner in 1791 and served in various bureaucratic capacities in Saxon mining. Humboldt corresponded with both Schelling and Ritter, and his call for a 'terrestrial physics' (*physique du monde*, *Physik der Erde*) also inspired Ferdinand Reich. In a 1797 work, Humboldt claimed to have attempted Thouvenel's pendulum experiments, to little effect.[70] The philosopher, Franx Xaver von Baader, studied three years under Werner before turning to *Naturphilosophie*. After taking a high position in Bavarian mining and metal works, as well as one in Bohemia, Baader became a philosophy professor at Munich. It was he who helped Ritter secure an appointment at the Bavarian Academy of Sciences in 1805. Finally, the physicist, Christian Samuel Weiss, was also a student of Werner's and his lifelong correspondent and friend,[71] and it was Weiss who first informed Ritter of the dowser Campetti. Weiss also promoted Ritter's work in Paris.[72]

That Freiberg knew of Ritter's work is confirmed in the work of the professor of *Bergbaukunst*, Moritz Ferdinand Gaetzschmann (1800–1895). A former Academy student and instructor of mine surveying, Gaetzschmann's textbook addressed both the most cutting-edge earth science and long-standing prospecting knowledge, as the author, not unlike Trebra before him, exposed a dual commitment to Academy and *Bergverständiger*. As Gaetzschmann explained, Step One in establishing a mine was prospecting. That included the "[natural] signs and

sie sich gründen, seit der Bekanntschaft mit dem Galvanismus und den Phänomenen des thierischen Magnetismus, freilich nicht ganz bezweifelt werden." Gotthilf Henrich von Schubert, *Handbuch der Geognosie und Bergbaukunde* (Nürnberg, 1813), 332.

[69] "Er [Ritter] war ein junger Mann von großem Talent, in der Chymie, auch in der Geschichte derselben wohl bewandert, und Kenntnisse, die ihm etwa noch fehlten, erwarb er sich mit Leichtigkeit." Henrich Steffens, *Was ich erlebte: aus der Erinnerung niedergeschrieben* (Stuttgart, 1995), 4:88.

[70] Alexander von Humboldt, *Versuche über die gereizte Muskel- und Nervenfaser, nebst Vermuthungen über den Chemischen Process des Lebens in der Thier- und Pflanzenwelt,* vol. 1 (Berlin, 1797), 470. On Humboldt's relation to Freiberg and mining, see especially Ulrike Leitner, ed., *Studia Fribergensia: Vorträge des Alexander-von-Humboldt-Kolloquiums in Freiberg von 8. Bis 10. November 1991 aus Anlaß des 200. Jahrestages von A.v. Humboldts Studienbeginn an der Bergakademie Freiberg* (Berlin: Akademie Verlag, 1994).

[71] *Allgemeine Deutsche Biographie*, s.v., "Weiß, Christian Samuel."

[72] Klengel, "Ultaviolette Strahlen," 256.

tools" in the discovery of minerals.[73] A thorough knowledge of rock formations, mineralogy, mineralogical chemistry, and physics formed the basis of prospecting for Professor Gaetzschmann. However, this "*Wissenschaft*" was compatible with a due appreciation for chance discovery and *Schürfen*. Gaetzschmann incorporated these beliefs and techniques too, referencing Johann Mathesius, Georg Agricola, Georg Löhneiß, and Balthasar Rößler. The independent miner analyzed the color of earths, springs and streams, he looked for particular herbs and plants, stunted growth and singed leaves, and other effects of warm mists (*Nebel*) and vapors (*Dünste*) from underground.[74]

Gaetzschmann then devoted some ten pages to the problem of dowsing, complete with visual depiction and a literature review. Like Trebra, Gaetzschmann remained skeptical but open-minded. "In earlier times," he recounted, "the *Wünschelruthe* stood in high esteem, and to some extent still today...it even found official application when dowsers [*Ruthengängern*] were paid, according to their claims."[75] Gaetzschmann presented all seven rods from the *Last Testament* of Basil Valentine (Thölde), a "preeminent authority among believers," who Gaetzschmann still believed had lived as a Benedictine monk in the fifteenth century.[76] But the scientist turned to the latest on electricity and galvanism: professional physicists balanced metal rods on the fingertips of receptive individuals, or hung pendulums from their fingers. A new language of electricity had come to replace the "sympathy of the old."[77] Sudden flashes of light in the mines (*Witterungen*) were merely electrochemical reactions. Gaetzschmann did remain skeptical on the matter of dowsing: How could galvanic current from a mineral vein transfer to the human body through a wooden stick? "Even if the dowser, following the proposal of one adept of about ten years ago, covered the soles of his feet and body with gold leaf?"[78] That man was undoubtedly one Karl Schmidt, who Ferdinand Reich examined in the 1840s. Still, Gaetzschmann allowed that dowsers possessed an inexplicable tacit knowledge of the mountain, and he recalled the older view that they detected mineral

[73] Moritz Ferdinand Gaetzschmann, *Einleitung zur Bergbaukunst* Part 1 (Freiberg, 1856), 253.
[74] Gaetzschmann, *Einleitung*, 287.
[75] Gaetzschmann, *Einleitung*, 295.
[76] Gaetzschmann, *Einleitung*, 302.
[77] Gaetzschmann, *Einleitung*, 305.
[78] Gaetzschmann, *Einleitung*, 306.

fumes: "It cannot be denied that many individuals of this sort, assisted by natural acumen, good powers of observation, even a peculiar physical receptivity to the action of moist vapors, have great value in their trade."[79]

Ferdinand Reich and the Schneeberg Dowser

Freiberg continued to show a formal interest in dowsing in Gaetzschmann's day, as dowsers could synthesize new science with *Schürfen* knowledge. Ferdinand Reich's investigation of Karl Wilhelm Schmidt has remained a quaint reference in the history of physics and Saxon mining, though it was a logical extension of Freiberg's broader interest in electrical theory.[80] When the Mining Office contacted Reich about Schmidt, the electromagnetic properties of mineral earth was one of the foremost problems in physics. In 1830, Robert Fox had demonstrated the phenomenon for copper ore using a Schweigger multiplier (or galvanometer).[81] Shortly after, the Office had asked Reich to examine earth with this new instrument, in part to determine whether it had any practical application in mining.[82] Following Fox's procedure, Reich confirmed the electrical or galvanic property of mineral rock, which he attributed to the interaction of fluid and minerals. As for

[79] Gaetzschmann, *Einleitung*, 334.

[80] Reich spent his career investigating physical 'forces' and their relevance to geology, but like Volta, Reich was no *Naturphilosoph* or supporter of Ritter. Reich worked with Ludwig Wilhelm Gilbert in 1815 at Leipzig on determining the density of the earth, and Gilbert, a close friend of Volta's, was an outspoken opponent of Ritter and *Naturphilosophie*. Gilbert published his opinions against Ritter in "Einige Kritische Aufsätze über die in München wieder erneuerten Versuche mit Schwefelkies-Pendeln, Wünschelruthen, u.d.m.," *Annalen der Physik* (1807), 2:369–449. See also Wagenbreth, *Die Technische Universität*, 88.

[81] Robert Were Fox, "On the Electromagnetic Properties of Metalliferous Veins in the Mines of Cornwall," in *Philosophical Transactions of the Royal Society* (1830). See Heinz Balmer, *Beiträge zur Geschichte der Erkenntnis des Erdmagnetismus* (Aarau, 1956).

[82] Ferdinand Reich, "Ueber electrische Stöme auf Erzgängen," *Jahrbuch für den Berg- und Hüttenmann* (Freiberg, 1840): 1. Reich would use two instruments with Schmidt. The electrometer contained two gold leaves suspended inside a jar. Since like charges repelled, when the experimentor introduced an electric charge, the identical leaves separated. Johann Schweigger devised another instrument (the *Multiplikator*, multiplier, or galvanometer) in 1820, which Reich also employed. Wire loops in a magnetic field increase the turning force of a current, observed by the deflection of a magnetized needle.

implications in mining, Reich insisted that that determination had to await further examination.[83]

Schmidt positioned his practice within the language of galvanism, and the physicist would use an electrometer and galvanometer to test the dowser, even if Reich rejected dowsing theory *a priori*. Schmidt was an Academy graduate (both he and Reich were enrolled in 1816, which suggests some prior acquaintance),[84] and a mine surveyor and shaft manager at Schneeberg since at least 1827. He recounted his experiences with the dowsing rod in a local mining journal in 1842, in which an elderly author, invoking Thouvenel and Ritter's work, spoke of "galvanic" responses and "electrical currents." A self-described "geognost," or student of geognosy, Schmidt evidently considered his dowsing practice to complement his earlier study at the Academy. He explained that misuse and deception had disgraced the otherwise valid practice of an "experienced [*Wissenschaftlichen*] miner."[85] Like Freiberg dowsers before him, Schmidt referred exclusively to the 'rod' (*Rute*, *Rutenschlag*, *Rutenschläger*), rather than 'divining-' or 'wishing rod' (*Wünschelrute*). He claimed that he was a disbeliever until becoming a mine surveyor, which brought him into contact with *Schürfen* and therefore with dowsing. Schmidt added that his own body was "receptive" (*empfänglich*).

Schmidt's self-experimentation led to two curious observations: that whenever he held the rod in the classic palm-up style, he felt "galvanic excitement," and that bulbous tree trunks seemed to grow exclusively above mineral earth. That claim was rooted in the longer-standing belief that mineral vapors affected growth aboveground. Mining books since the sixteenth century had developed the notion, though Schmidt may have been substituting electrical impulses for vapors. The telltale bulbs did not parallel underground veins exactly, Schmidt conceded, but tended to follow their trajectory, and dense groupings indicated an intersection of veins.

---

[83] "...so bleibt die genauere Ermittlung dieses für den Bergmann wichtigsten Umstandes noch weitern Versuchen vorbehalten." Reich, "Ueber electrische Ströme," 2.

[84] *Festschrift zum hundertjährigen Jubiläum der Königl. Sächs. Bergakademie zu Freiberg am 30. Juli 1866* (Dresden, 1866), 248.

[85] "Haben Mißbrauch und Betrug den Ruthenschlag gleichsam infam und aller nähern Untersuchung als eines wissenschaftlichen Bergmannes unwürdig, lächerlich gemacht, so muß der Wahrheit doch die Ehre gegeben werden." *Notizen über den Rutenschlag*, in August Friedrich Wappler, "Alte sächsische Wünschelruten-Geschichten," *Mitteilungen des Freiberger Altertumsvereins* 43 (1907): 77.

Schmidt described a dowsing method that was highly systematic, and which closely resembled the practice as Rößler had explained and depicted in his text of 1700 [Fig. 7]. Having prepared stakes of various sizes, Schmidt proceeded to a prospective field and walked with his dowsing rod in straight lines, charting evenly spaced and parallel paths. Wherever the rod responded, an assistant would hammer in a stake. Schmidt then crisscrossed the parallel lines at right angles with new lines, forming a grid. The assistant hammered in new stakes. Then, Schmidt intersected the angles by pacing diagonal paths. Once the entire region was thus staked, Schmidt zeroed-in on highly staked-out regions in more circular fashion to rule out "false impulses." The entire process exhausted his body, since the strength of electrical currents increased over time.[86]

The Office invited Schmidt to Freiberg for examination by Professor Reich. Schmidt's only concern was that cost considerations not hinder the examination, since a fully staked-out field would require multiple digs.[87] We also learn that Schmidt's examiners would choose suitable ground on which to test him—it would be level, known to be rich, and have no visible indication of mineral ore—and the men would keep their plans hidden from Schmidt. Reich and his team were thorough: since earth in the vicinity of working mines was likely to contain mineral fragments wherever one digged, irrespective of a dowser's presumed ability, Schmidt would have to distinguish between proper veins and mere mineral fragments.

The first trial took place on October 12, 1843. According to Reich's report, the men walked to a pre-selected area next to a stable that was unfamiliar to Schmidt. He then proceeded to dowse the region in exactly the same manner he had outlined in his article. He walked with "small steps slowly forward," carrying the instrument Reich termed his "*Wünschelruthe*" before him, an apparatus consisting of intertwined iron wires, fixed into a metal cap, fitted with brass grips, and covered in a mesh of copper wire. That Schmidt also coated his shoes in gold leaf (as Gaetzschmann later recalled) underscores how central an electrical or galvanic theory was to these experiments, as the metallic rod and

---

[86] "Läuft ein Ruthengänger aber gleich anfänglich continuirlich auf einer Gangregion fort, so impulsiren die elekrischen Ströme seinen Körper nach und nach so steigend, daß seine Schritte endlich unzuverlässig werden und das kleinste übersetzende Klüftchen ihn auf Abwege führt." Wappler, "Wünschelruten-Geschichten," 78.

[87] BAF, Nr. 2911, Bl. 55.

the gold-leafed shoes were intended to conduct mineral current. That the dowser had another rod made of fish bone and wooden grips suggests, however, that his own theory was more expansive than that of the physicists.

Schmidt claimed to have identified four veins that day, the presumed intersections of which marked four later digs. Mine surveyor Leschner laid the string and used his compass to chart the direction, strike, and intersections of Schmidt's claims. Excavation of the four sites took place on October 20, and the results confirmed Reich's suspicion that one could find mineral earth almost anywhere around Freiberg. The first dig [Fig. 9] passed through topsoil (*Dammerde*), then a rough layer packed with sharp-edged stones, and then a thin band of iron-bearing earth containing large fragments of quartz and bloodstone (*Roteisenstein*). At bottom was harder metamorphic rock (gneiss) with thin fissures (*Klüfte*). The second dig, which required that the diggers leave a buffer (*Strosse*) against encroaching waters, resembled the first dig in all respects except the base. There, the diggers recognized thin veins (*Trümmer*) running through the gneiss in addition to fissures, and at the corner, a larger muddy vein of iron-bearing gneiss rock. The third and fourth digs produced a similar stratification, with thin quartz-bearing veins appearing in the third. Schmidt supposed that the dark layer with large mineral fragments (the third layer down) originated from a substantial mineral vein nearby, from which the iron-bearing vein discovered in Dig 2 branched. In his sketches, he accordingly placed Dig 2 to the side of the other three columns, and stated in the caption below that the iron earth, "may well have originated from a nearby, strong vein." He added that this earth had a "strong effect on the rod." Reich reported Schmidt's interpretation, but specified in his footnotes that the party had not found the supposed veins or intersections about which Schmidt had originally spoken.[88]

Reich was inclined to discount the dowser, and further experiments on October 20 seemed to confirm the skepticism. Schmidt was asked to dowse a pathway (labeled as *Bauergasse* at left on Leschner's map [Fig. 10]; today Hainichenstrasse) under which veins were known to pass. The spots at which Schmidt's rod dipped strongly are labeled as 'No 1' 'No 2' and continuing. Reich considered these results strong proof against Schmidt, since his rod did not locate either the Priestly

[88] Wappler, "Wünschelruten-Geschichten," 81–82.

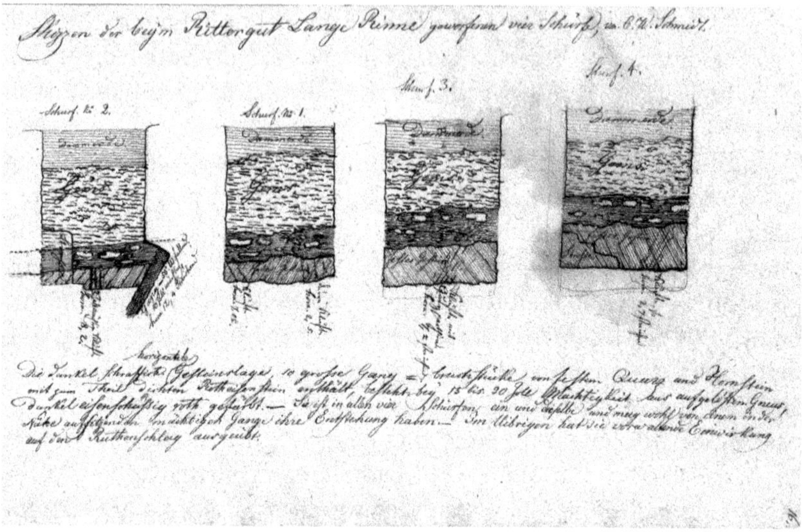

Fig. 9. The dowser Karl Schmidt's sketch of four digs (*Schürfe*). The dark band containing chunks of mineral rock "exerted a strong effect on the rod's dip [Ruthenschlag]."

Felicitation (*Priesterliche Glückwunsch*) or the Maria and Max veins, both marked on the map as red lines crossing Bauergasse. Schmidt explained this oversight in a letter of October 21. Addressing the "Dowsing Committee," he claimed that had he paced the region more thoroughly, he would certainly have experienced the effects of these veins.[89] He also referred to his published essay that the committee had read (and obviously valued), in which he conceded that the dowsing rod often dipped in "various" and "inexplicable" ways.[90] This was evidence of the delicacy of the art rather than its dubiousness.

But Schmidt had a more sophisticated defense. As Leschner's map shows, miners had discovered any number of offshoots from the Jonas, Esselstollen, and Michaelis veins shown at right. They are labeled with 'A' 'B' 'C' and continuing. The final letter, 'H', appears to the left, beside a group of branches off of Priestly Felicitation. Schmidt believed that these tributary veins crisscrossed the area according to the light dotted lines indicated. In particular, he claimed that the spots along the

[89] BAF, Nr. 2249, Bl. 20.

[90] "... bey dem Ruthenschlag Vorkommnisse stattfinden, die örtlich vielseitig und unerklärbar sich gestalten," BAF, Nr. 2249, Bl. 20b.

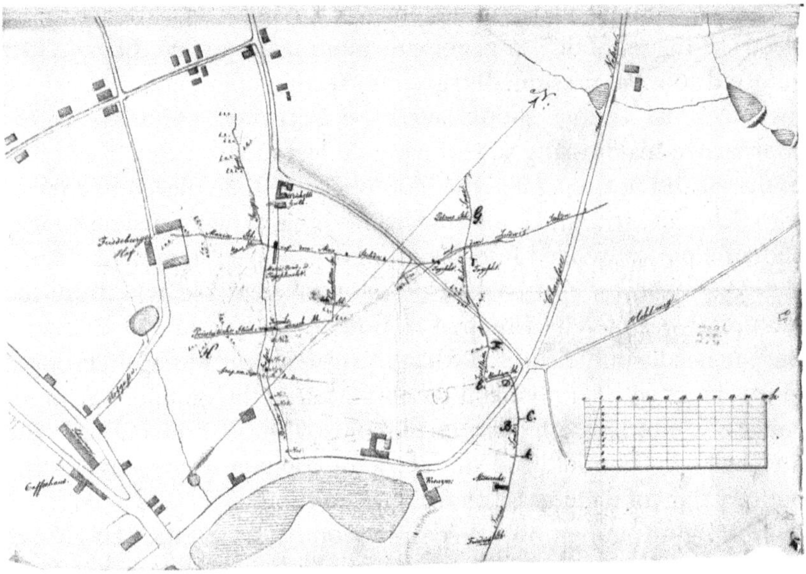

Fig. 10. The mine surveyor produced a map of mineral veins, pathways, and structures on land dowsed by Karl Schmidt. The dotted lines represent uncharted vines, according to the dowser.

pathway over which his dowsing rod dipped—proof for Reich that it dipped falsely—were the places at which several of these uncharted veins continued below, or intersected with others. So, for example, a small branch off the Esselstollen vein, labeled with 'F' to the right, crossed the land and reached the pathway at No. 2, where Schmidt had placed a stake. Another vein from 'A' also intersected here at No. 2, as did a third from Priestly Felicitation above. Schmidt apologized for not making these deductions with the party the day before. Had he been privy to Leschner's chart, Schmidt explained, he would have realized that the rod, while it failed to locate two major veins, dipped to important sources nevertheless.[91] The correspondences between stakes and veins was "most likely," he believed, and "it would be difficult not to acknowledge this most decisive evidence."[92] These remarks did not survive into Reich's official report.

[91] BAF, Nr. 2249, Bl. 21.

[92] "…es wohl schwer seyn möchte diesen bescheid etlichst ausgesprechenen Nachweis nicht anerkennen zu wollen." BAF, Nr. 2249, Bl. 21.

More damning still, Schmidt's practice did not lend credence to the electrical theory of dowsing, notwithstanding his golden shoes. When he stood above a (presumed) triple intersection of mineral veins where the electrical charge should have been strong, the investigators approached his dowsing rod with a gold leaf electrometer. The leaves in the jar did not separate. The multiplier was then placed in contact with Schmidt's rod, and it too showed nothing. When tested on copper and zinc pieces wedged into the ground, however, it showed a small charge as expected.[93] At stake Nr. 5 along Bauergasse, wire from the multiplier was literally wrapped around Schmidt's rod and observed. But the needle only deflected when the instrument was again brought into contact with damp metal pieces placed in the ground.[94] Schmidt could also not locate hidden metal coins with his dowsing rod. The men brought Schmidt inside the academy complex itself (shown at the bottom-right of Bauergasse on the map) to conduct the classic coin test that brought Campetti to Ritter's attention back in 1805. The dowser would locate coins that Reich, Leschner, and the Bergmeister from Marienberg (later *Oberberghauptmann*), Friedrich Constantin Freiherr von Beust, had hidden in advance.[95] Schmidt had confirmed that his rod dipped to piles of metal coin. The men laid 17 chests on the floor and filled one (Number Two) with one- and two-taler pieces. Schmidt was let in and began wandering the chests with his rod. He recognized an effect at Number Three and Number Eleven, deciding finally on Eleven, which was empty. He then left the room and the gentlemen arranged a second test, placing the coins in Number Six. Schmidt returned, paced the chests, and decided first on Fourteen, then on Seven, then on Eight, and finally on Number Nine, the wrong chest again.

Still confident in the authenticity of his rod despite failing the coin test, Schmidt invited Beust to grasp the rod and be led by Schmidt, who grabbed from the front and walked backwards. That produced "no decisive result," said Reich, because Beust was uncertain whether the rod or Schmidt himself caused the motions that Beust felt.[96] One

---

93 Wappler, "Wünschelruten-Geschichten," 80.

94 Wappler, "Wünschelruten-Geschichten," 83.

95 Wappler, "Wünschelruten-Geschichten," 82. Beust (1806–1891) had studied at Freiberg, Leipzig, and Göttingen, and worked in different Saxon *Bergämter*, before being named *Oberberghauptmann* in 1851. His brother became prime minister of Austria-Hungary.

96 "…kein entscheidendes Resultat, indem der Unterzeichnete [Beust] sich nicht zu überzeugen vermogte, ob wenn eine Bewegung erfolgte, dieselbe von der Ruthe

might imagine these two men grasping the rod together, face to face, and inclined toward competing interpretations of the same dipping motions, like players of a modern-day ouija board.

Reich's report concluded that there was "not a single fact" to suggest indubitably that Schmidt's rod determined the location of mineral veins, although he felt "obliged" to add that the dowser remained convinced.[97] Schmidt and the examiners, predisposed toward opposing interpretations, had a great difference of opinion when observing complex data. Schmidt vaunted the subtlety of his art and formed his explanations to some degree post hoc, whereas Reich looked for unambiguous experimental results. Schmidt possessed a dowsing practice that synthesized scientific theory and his tacit knowledge, and Reich upheld an electrical theory that excluded the experience and interpretations of the dowser. In 1846, Schmidt reported new experiences with his dowsing rod to the Mining Office, but this time Reich refused to investigate.[98]

Others would. The archival record shows that Freiberg continued to conduct experiments with presumed human electrometers well into the twentieth century.[99] Had Schmidt wished to press his case further, he might have cited Dietrich Georg Kieser's article of 1819 in the *Archive for Animal Magnetism*, in which the author was adamant that the sidereal force and animal magnetism were unlike magnetism and electricity, and did not register on an electrometer.[100]

ausgieng oder durch Herrn Markscheider Schmidt veranlaßt werde." Wappler, "Wünschelruten-Geschichten," 83.

[97] Wappler, "Wünschelruten-Geschichten," 83–4.

[98] Täschner, "Ferdinand Reich," 41.

[99] BAF, 40024-16, Nr. 45, "Wünschelrutenforschungen," 1920. See also Hubert Knoblauch, *Die Welt der Wünschelrutengänger und Pendler: Erkundungen einer verborgenen Wirklichkeit* (New York: Campus Verlag, 1991).

[100] Kieser became President of the German Academy of Natural Sciences (Leopoldina) in 1858. The mineral kieserite is named after him.

CONCLUSION

In an interesting piece for the journal, *Social Studies of Science*, Susan Schmidt Horning argues that recording engineers of the twentieth century, notwithstanding great advances in the technology of recording, relied on tacit knowledge in the use of recording horns and microphones. As the studio environment became more technological, acoustical recordists, working directly with musicians and singers, integrated their technical training with the 'art' of recording. How to position and operate a microphone in a high-tech environment was never learned in manuals or the classroom.[1] This resembles prospecting practice as we have described it: the recordist had a feel for the way equipment captured sound, much like the *Bergverständiger* had a tacit knowledge of how the divining rod detected mineral fumes or currents. In both cases, the practitioner integrated his idiosyncratic practice into an increasingly bureaucratized and technologized industry.

Horning's paper illustrates that the concept of tacit knowledge has application in surprising contexts. Historians of science have studied the artisanal dimensions of natural inquires, including humanist and philosophical interest in the crafts, but they might also reconsider Polanyi's concept. We have extended it to account for divining practice, and other researchers may wish to approach better-known magical practices in this light. When the astrologer drew up a nativity and zeroed-in on a client's life experiences or medical concerns; when the alchemist mixed his substances and worked his coals in the furnace; and when the sorcerer or soothsayer focused her attention on a reflective object and mediated information to her clients, they were activating a reservoir of unspoken knowledge, hunches, and intuition. Properly speaking, such individuals did not *learn* their respective arts so much as realize that they had extraordinary sensitivities and intuitions. Frauds and swindlers abounded, to be sure, but focusing on tacit knowledge allows us to better understand the 'true' magicians, as it were. Men like Karl Schmidt remained convinced until the end, even earning the respect of hard-nosed empiricists like Ferdinand Reich.

[1] Susan Schmidt Horning, "Engineering the Performance: Recording Engineers, Tacit Knowledge, and the Art of Controlling Sound," *Social Studies of Science* 34:5 (2004), 703–731.

This study was also an exercise in symmetrical analysis. The professionalization of earth science marked the birth of amateur science, as the transformation of discourses in science and divining, respectively, paralleled and reflected each other. Since at least the days of Georg Agricola and Christoph von Schönberg, with whose portrait we began, mining experts and scholars of the earth had flocked to Saxony. Traditions of mineralogy, physical geography, and advanced construction took root, trends that eventually culminated in the Freiberg Mining Academy. Historians of mining and earth science have emphasized progress and modernization in Saxon mining. But individual prospectors also flocked to Freiberg, dowsers among them, offering mining administrations valuable knowledge about the location of mineral resources. This too was a trend that culminated in the eighteenth century, when major officials employed *Bergverständige* like Christoph Dietrich, and when a modern electrical theory came to displace mineral vapor theory. Dowsers after Karl Schmidt increasingly spoke the language of electrical currents and receptivity, and they more frequently carried metallic rods into the field than in prior centuries.[2]

Recent studies of vernacular and artisanal knowledge in early-modern Europe, while they challenge our assumptions about science and bring new voices to the discussion, do not always introduce novel social and cultural contexts. More often than not, we find ourselves on well-trodden and familiar terrain: the streets of Seville or London, never too far from the famous courts, museums, universities, or academies of Europe. Preoccupation with high profile artisans, alchemists, humanists, physicians, and natural philosophers from Paracelsus and Dürer, to Becher, Bacon, and Newton, still drives the analysis. These are the "plains" about which Braudel spoke in the passage that opened our study. Telling our tale required that we leave the plains and comfortable bustle of city life, "approach the high mountains," and descend into the unfamiliar world of Saxon mining. This meant spanning the social spectrum to a greater degree than many others have, describing a mining culture that was common to diverse personalities: treasure hunters and diggers, local officials in the field, mining bureaucrats in Dresden, and even the Elector himself. We addressed scholarly theory from Agricola to Werner and Reich, but also presented microhistories of dowsing practice at Gommlo and Freiberg.

[2] C. A. Browne, "Observations upon the Use of the Divining Rod in Germany," *Science* 73:1882 (1931), 84–6.

There was no essential decline of Renaissance-era magic in this case, which historians have found for alchemy and astrology during the Enlightenment. As particular elements of folk experience ascended the social ladder, they elicited either encouragement or dismissal by scholars, theologians, and officials; other beliefs descended from the level of learned authorities to be adapted or rejected by foremen, diggers, and treasure hunters. The transmission of ideas either upwards or downwards was sometimes short-lived: clerics since the days of Father Mathesius rejected the mining spirits observed by diggers, whereas mining officials like Löwel and Beyer rejected corpuscular and other mechanical theories to explain dowsing. In other instances, the dialogue between high and low cultures was more fluid. Valentine (Thölde) first explained how dowsing rods responded to mineral vapors, and both rods and vapors appeared in mining folklore, mining books, and scholarly treatises into the eighteenth century. Early physicists proposed an electrical theory of dowsing, though mine surveyor Karl Schmidt, who otherwise appealed to galvanism, first encountered dowsing through *Schürfen*, and his practice resembled that explained in earlier mining books.

Whether the divining rod 'really worked' was never at issue here. While it seemed to me that a dowser pacing the Dübener Moor in 1713 was bound to find brine, and that a practitioner walking the Ore Mountains in the 1730s would eventually find mineral-bearing earth, irrespective of the practices involved, I also became convinced that dowsers, using their common sense, mining experience, and feel for the terrain were especially adept at finding these resources. The dowsing rod may have helped focus the practitioner's tacit knowledge of the mountain, even sharpening it, much like throwing a stone might concentrate and heighten one's anger. I therefore sympathize less with Ferdinand Reich than with Moritz Ferdinand Gaetzschmann, who could summarize cutting-edge science in his textbook, but allow that dowsers had "natural acumen."[3] But we leave it to cognitive psychologists and philosophers of mind to explore human knowledge and mental capacities. We end content for having explained what folklorists have known all along: that rational men in mining, involved in the professionalization and modernization of their industry, could follow a man with a magic wand through the fields.

[3] Gaetzschmann, *Einleitung*, 334.

SELECTED BIBLIOGRAPHY

Primary Sources

Unprinted Sources

Sächsisches Haupstaatsarchiv Dresden

Geheimes Finanzkollegium, Bergwerkssachen:

—— Loc. 36070 Bericht, von denen Ertz-Gebürgen, Streichenderer Gänge, Stöcke, Flöze, Klüffte, Ertze, Berg Arthen und allen Metallen, auch von Schürffen, Seiffenwercken und andern Arthen der Bergwercken, 1601.

—— Loc. 36263 Acta, Die Anehmung u. Verpflichtung Tobias Häußlers zum Rüthenganger u. dessen ausgesetstes jahrl: Wartegeld. 1739–1743.

—— Loc. 36302 Acta, den von des Herzogs zu Weymar Durchl: auf einige einige Zeit verlangten Freibergischen Ruthen-Gänger, Dietrich bet., 1743.

—— Loc. 25192 Cammer Acta, Die von Johann Friedrich Junghansen zu Rahniß vorhabende Aufsuchung einiger unterirrdischen Sachen betr, 1747.

—— Loc. 36062 Ein Heft betitelt: gründliche Bergbelustigung, enthaltend eine genaue Anweisung, mit der Wünschelrute Erzgänge aufzufinden, 1668.

Geheimes Archiv, Alchemische Sachen:

—— Loc. 4486 Alchimistischen Sachen, 1583–1708 in specie Projecte und Processe im Schmelz, Probirn, und Scheidesachen.

—— Loc. 4416/06 Verschiedenen Alchymistische Schriften, 1583.

—— Loc. 4418/1-2 Zwey unterschidene Bucher, worinnen schreiben, damit Churfurst Augustus und Churfurst Christian zu Sachsen sich allerhand Kunstler mit ihrer erfundenen Kunststücken angegeben von bei 1558 bis auf das 1590 Jahr...sonderlich in Alchemischen, Kriegen, Bergwercken, Wasser, Holtz, Salz, Munz, und Mathematischen sachen.

—— Loc. 7291/02 Acta verschiedene vorgeblicher Bewirkung wichtiger Vortheile in Vorschlag gebrachte Projecti, Maschinen und Münzoperationen auch andere Erfindungen und deren Unterstützung mit Gelde und sonst, ingleichen Entdeckung verborgener Schätze betr. Vom Jahr 1700. bis 1729.

—— Loc. 10690/47 Schatzgräberei, Hexerei, Wunderzeichen 1699, 1693–1735.

Sächsische Landesbibliothek, Staats- und Universitätsbibliothek Dresden

—— Schrank B, Nr. 158. Memorial, die Bergwissenschaft betreffend. 1671. Abraham Löwel, Bergmeister.

—— Schrank N, Nr. 87. Der rechte Geistliche Gezwang der Höllen und Schatzgräber Kunst.

—— Schrank N, Nr.165b, Bl.16–35. Johannis Faustus Höllen-Zwang.

—— Schrank N, Nr. 166. Speculum Salominis.

—— Schrank R, Nr. 284. Beschreibung des Edlen Bergwerckbaues, wie solcher bey der Churfürstl: Sächsch: alten Freyen Berg Stadt Freyberg verführet worden und noch verführet wird. Allen denenjenigen so sich bey demselben encapiret machen wollen sehr dienlich. verführet worden.

Bergarchiv Freiberg
——Loc. 40013, Nr. 174. Berichten, Salz Quelle in der Dübische Heyde.
——Loc. 40014, Nr. 74. Leipzig. Bericht über die Tätigkeit... Häusler.
——Loc. 40001, Nr. 2911 F00507 Ueber electrische Stömungen.
——Loc. 40001, Nr. 2249. Acta die Wünschelruthe betr., F. Reich [verfilmt]
——Loc. Nr. 2092. Die Bestellung und Verschickung derer Ruthengänger, ingl. deren Wartegeld betr., Bergtamt Freyberg an. 1728.
——Loc. 40024-16, Nr. 45 Wünschelrutenforschungen, 1920.

Landes- und Forschungsbibliothek Gotha
Katalog der occidentalischen Handschriften:
——Ch. B., Nr. 1432. Die virgula divina, der Rüthen Buch.
——Ch. A., Nr. 972. Unterricht vom Bergbau, vornemlich in der Mackscheiderkunst. Augustus Beyer. 1718.

Printed Primary Sources

Agricola, Georg. *Zwölf Bücher vom Berg- und Hüttenwesen*. München: Deutscher Taschenbuch Verlag, 1994.
Albinus, Petrus. *Meißnische Bergk Chronica*. Dresden, 1590.
Anonymous. *Das ABC der Bergwercks-Wissenschaften, oder Kurze und deutliche Anfangs-Gründe der Erkenntniß des Bergwercks in Frage und Antwort abgefasset zum Unterricht für die Jugend welche diesen Theil der Natur-Lehre erlernen will*. Freiberg, 1747.
Bergcalendar. Gotha Forschungsbibliothek, Math 8°, 1260, 1773–1784.
Beyer, Augustus. *Grundlicher Unterricht von Bergbau, nach Anleitung der Marckscheiderkünst, worinnen geweisen wird, wie die Gebürge und darinnen streichende Gänge, Klüffte und Flöze zu untersuchen und zu beurtheilen*. Freiberg, 1749.
Biringuccio, Vannoccio. *Pirotechnia*. Translated by C. S. Smith M. T. Gnudi. New York: Dover Publications, 1959; orig. publ. 1942.
Charpentier, Johann Friedrich Wilhelm von. *Beobachtungen über die Lagerstätte der Erze, Haupsächlich aus den Sächsischen Gebirgen*. Leipzig, 1799.
Ercker, Lazarus. *Aula Subterranea domina dominatium subdita subditorum. Das ist, Untererdische Hoffhaltung, ohne welche weder Herren regieren, noch die Unterthanen gehorchen können. Oder Gründliche Beschreibung der jenigen Sachen, so in der tieffe der Erden wachsen*. Prague, 1574.
Feudiviro [Christian Gottlieb Lehmann]. *Gebrauch der Berg- und Wünschel-Ruthe, was und wie vielerley sie sey, woraus sie gemacht, und was die Ursache ihres Schlags auf Erz-Gängen, Wasser, todte Körper und Schätze sey, sowohl derselben Mißbrauch*. Leipzig, 1763.
Gaetzschmann, Moritz Ferdinand. *Einleitung zur Bergbaukunst* Part 1. Freiberg, 1856.
Henckel, Johann F. *Unterricht von der Mineralogie oder Wißenschafft von Waßern, Erdsäfften, Saltzen, Erden, Steinen und Ertzen, nebst angefügten Unterricht von Chymia Metallurgica, wie selbigen der Wohlseel. Herr Berg-Rath Joh. Friedrich Henckel, sowohl seinen in der Mineralogie und Chymie gehabten Scholaren discursive ertheilet, als auch der Nachwelt zum Dienst in Manuscripto hinterlaßen*. Dresden: Joh. Nicol. Gerlachen, 1747.
——. *Flora Saturnizans, die Verwandschaft des Pflanzen mit dem Mineralreich, nach der Naturalhistorie und Chymie aus vielen Anmerkungen und Proben*. Leipzig, 1755.
——. *Mineralogische, Chemische und Alchymische Briefe von reisenden und andern Gelehrten an den ehemaligen Chursächsischen Bergrath J.F. Henkel* (Dresden, 1794).

Jugel, Johann Gottfried. *Geometria Subterranea, oder Unterirrdische Meßkunst der Berg- und Grubengebäude, sonst insgemein die Marckscheidekkunst genannt.* Leipzig, 1773.

Kircher, Athanasius. *Magnes sive de arte magnetica opus tripartitum quo praeterquam quod universa magnetis natura, eiusque in omnibus artibus & scientiis usus noua methodo explicetur, e viribus quoque & prodigiosis effectibus magneticarum, aliarumque abditarum naturae motionum in elementis, lapidibus, plantis & animalibus elucescentium, multa hucusque incognita naturae arcane per physica, medica, chymica, & mathematica omnis generis experimenta recluduntur.* Rome, 1641.

——. *Mundus Subterraneus, in XII libros digestus; quo divinum subterrestris mundi opificium, mira ergasteriorum naturae in eo distributio, verbo pantamorphou Protei regnum, universae denique naturae majestas & divitiae summa rerum varietate exponuntur. Abditorum effectuum causae acri indagine inquisitae demonstrantur; cognitae per artis & naturae conjugium ad humanae vitae necessarium usum vario experimentorum apparatus, necnon novo modo, & ratione applicantur.* Amsterdam, 1665.

Lehmann, Christian. *Ausführliche Beschreibung des Meißnischen Ober-Ertzgebürges, nach seiner Lage, Gestalt, Bergen, Thälern, Felsen, Flüssen, Brunnen, warmen Bädern, Wäldern, Lands-Art, Früchten, Wildbahne, wie auch angemerckten Zustande der Elemente, Himmels-Zeichen, Witterung und allerhand curiosen Begebenheiten gefertigt, auch mit schönen Kupffern und nöthigen Figuren gezieret von Einem Freunde des Ertzgebürges.* Leipzig, 1747.

Lehmann, Ernst. *Georg Agrikola's Mineralogische Schriften.* Freiberg: Craz and Gerlach, 1812.

Lehmann, Johann G. *Abhandlung von den Metall-Muttern und der Erzeugung der Metalle aus der Naturlehre und Bergwerckswissenschaft, hergleitet und mit chymischen Versuchen.* Berlin, 1753.

Leibniz, Gottfried W. *Protogaea.* W.E. Peuckert, ed. Stuttgart: W. Kohlhammer, 1949.

Löhneyß, Georg E. von. *Bericht vom Bergkwerck, wie man dieselben Bawen und in guten Wolstandt bringen soll; sampt allen darzu gehörigen Arbeiten, Ordnung und rechtlichen Proceß.* Zellerfeldt, 1617.

Magnus, Albertus. *Book of Minerals*, Wyckoff, D., tr. Oxford: Clarendon Press, 1967.

Martini, Johann Leonhard. *Brieffe oder Send-Schreiben, Vornehmer und Gelehrter Leute, welche die Verspottung der Wünschel-Ruthe vorstellen, und deren Systemata oder Grund-Sätze über einen Haussen werffen.* 1700.

Mathesius, Johann. *Sarepta oder Bergpostill.* Nuremberg, 1564.

Montanus, Elias. *Ein Büchlein von dem Bergwergk, wie man dasselbige nach der Rutten unnd Witterung bawen sol, allen so darzu lust haben, sehr dienstlich und zu willen nötig.* Zerbst, 1600.

Oppel, Christian F. W. von. *Anleitung zur Markscheidekunst nach ihren Anfangsgründen und Ausübungen kürtzlich entworfen.* Dresden, 1749.

Otto, Andreas, ed. *Der heimliche und unerforschliche Natur-Kündiger, oder: Accurate Beschreibung von der Wünschel-Ruthe, darinnen enthalten der besondere Nutz bey Entdeckung der Wasser-Quellen, Metallen, vergrabenen Schätze, flüchtiger Diebe und Mörder…Auch mit einem absonderlichen curieusen Tractat Mathia Willen, von der Wünschel-Ruthen vermehret.* Nürnberg, 1694.

Praetorius, Johann. *Gazophylaci Gaudium, das ist Ein Ausbund von Wündschel-Ruthen oder sehr lustreiche, und ergetzliche Historien von wunderseltzamen Erfindungen der Schätze, so geschehen seynd.* Leipzig, 1667.

Prescher, Hans, ed. *Georgius Agricola—Ausgewählte Werke III: Schriften zur Geologie und Mineralogie I.* Berlin: Deutscher Verlag der Wissenschaften, 1956.

Ritter, Johann W., ed. *Der Siderismus* I. Tübingen, 1808.

Rößler, Balthasar. *Speculum Metallurgiae Politissimum. Oder Hell-polierter Berg-Bau-Spiegel, darinnen zu befinden: wie man Bergwerck suchen, ausschürffen, mit Nutzen bauen allenthalben wohl anstellen, beförden, dabey alles Gestein und Ertze gewinnen, fördern, rösten, schmelzen und zu gut machen, dann auch was darbey zu thum oder zu lassen, hierüber ein iedweder, so dem Bergwerck zugethan ist, wissen und verstehen soll*. Dresden, 1700.

Schaffrath, Wilhelm M. *Codex Saxonicus: chronologische Sammlung der gesammten praktisch-gültigen Königlich Sächsischen Gesetze von den ältesten Zeiten, vom Jahre 1255 an bis zum Schlusse des Jahre 1840*. Leipzig: Reclam, 1842.

Schönberg, Abraham von. *Ausführliche Berg-Information, Zur dienlichen Nachricht vor Alle, die Bey dem Berg und Schmelzwesen zu schaffen*. Leipzig, 1693.

Sperling, Johann. *An Virgula mercurialis agat ex Occulta Qualitate, Disquisitio Philosophica*. Wittenberg, 1666.

Sudhoff, Karl. *Paracelsus: Sämtliche Werke. I Abteilung. Medizinische, naturwissenschaftliche und philosophische Schriften Vol. I–XIV*, 1922–1933.

Thomasius, Christian. *Versuch von Wesen des Geistes oder Grund-Lehren, so wohl zur natürlichen Wissenschafft als der Sitten-Lehre. In welchen gezeiget wird, daß Licht und Lufft ein geistiges Wesen sey, und alle Cörper aus Materie und Geist bestehen, auch in der gantzen Natur eine anziehende Krafft, in dem Menschen aber ein zweyfacher gutter und böser Geist sey* (Halle, 1699).

Trebra, Friedrich W. H. von. *Bergmeister-Leben und Wirken in Marienberg, vom 1. Decbr. 1767. bis Augusti 1779*. Leipzig: VEB Deutscher Verlag für Grundstoff industrie, 1990.

——. *Erfahrung vom Innern der Gebirge, nach Beobachtungen gesammlet*. Leipzig, 1785.

Valentine, Basil [Johann Thölde]. *Geheime Bücher, oder Letztes Testament. Vom grossen Stein der Uralten Weisen und andern verborgenen Geheimnussen der Natur. Auss dem Original, so in dem hohen Altar zu Erffurt unter einem Marmorsteinen Täfflein gefunden, nachgeschrieben*. Strasbourg, 1645.

Voigtel, Nicolaus. *Geometria Subterranea, oder Marckscheide-Kunst, darinnen gelehret wird wie auff begwercken alle Klüffte und Gänge in Grund und am Tag gebracht, auch solche von einander unterschieden werden sollen*. Eisleben, 1686.

Werner, Abraham. *Neue Theorie von der Entstehung der Gänge, mit Anwendung auf den Bergbau, besonders den freibergischen*. Freiberg: Gerlach, 1791.

Zeidler, Johann G. *Pantomysterium, oder das neue vom Jahre in der Wündschelruthe als einem allgemeinen Werckzeuge menschlicher verborgenen Wissenschafft. Darinnen ihre Natur und Ursach ihrer Bew… auch weitlauffiger Gebrauch und Nutz aus des Autoris eigener Erfahrung, und Physicalischen … philosophisch ausgeführet, und der gelehrten Welt zu weitern Nachsinnen übergeben wird*. Hall in Magdeburg, 1700.

Zimmerman, Christoph F. *Ober-Sächsische Berg-Academie, in welcher die Bergwercks-Wissenschaften nach ihrem Zusammenhange entworffen werden. Alles aus historischen Nachrichten, gründlichen Untersuchungen, natürlichen Beobachtungen, chymischen und mechanischen Versuchen, und darbey vorgefallnen Anmerckungen erläutert*. Dresden and Leipzig: Friedrich Hekel, 1746.

Secondary Sources

Antognazza, Maria R. *Leibniz: An Intellectual Biography*. Cambridge, NY: Cambridge University Press, 2009.

Arnt, Adolph. *Zur Geschichte und Theorie des Bergregals und der Bergfreiheit*. Halle: C.E.M. Pfeffer, 1879.

Ash, Eric H. *Power, Knowledge, and Expertise in Elizabethan England*. Baltimore: The Johns Hopkins University Press, 2004.

Bächtold-Stäubli, Hanns. *Handwörterbuch des deutschen Aberglaubens*. Berlin and Leipzig: W. de Gruyter & Co., 1927–1942.

Barrett, William, and Besterman, Theodore. *The Divining Rod: An Experimental and Psychological Investigation*. New York: University Books, 1968.

Bartels, Christoph, and Denzel, Markus A., trans. *Konjunkturen im Europäischen Bergbau in vorindustrieller Zeit. Festschrift für Ekkehard Westermann zum 60. Geburtstag*. Stuttgart: Franz Steiner Verlag, 2000.

Baumgärtel, Hans. *Vom Bergbüchlein zur Bergakademie: Zur Entstehung der Bergbauwissenschaften zwischen 1500 und 1765/1770*. Leipzig: VEB Deutscher Verlag für Grundstoffindustrie, 1965.

——. *Bergbau und Absolutismus: Der sächsische Bergbau in der zweiten Hälfte des 18. Jahrhunderts und Maßnahmen zu seiner Verbesserung nach dem Siebenjährigen Kriege*. Leipzig: VEB Deutscher Verlag für Grundstoffindustrie, 1963.

Behringer, Wolfgang. *Hexenverfolgung im Bayern: Volksmagie, Glaubenseifer und Staatsrason in der Fruhen Neuzeit*. Munchen: R. Oldenbourg, 1988.

Beretta, Marco. "Humanism and Chemistry: The Spread of Georgius Agricola's Metallurgical Writings." *Nuncius* 12 (1997).

Beringer, Carl C. *Geschichte der Geologie und des Geologischen Weltbildes*. Stuttgart: Ferdinand Enke, 1954.

Bernardi, Walter. "The Controversy on Animal Electricity in Eighteenth-Century Italy: Galvani, Volta, and Others." *Nova Voltiana: Studies on Volta and his Times* I (2000).

Bird, Christopher. *The Divining Hand: The Art of Searching for Water, Oil, Minerals, and other Natural Resources or anything Lost, Missing, or Badly Needed*. New York: E.P. Dutton, 1979.

Blanchard, Ian. *International Lead Production and Trade in the 'Age of the Saigerprozess' 1460–1560*. Stuttgart: Franz Steiner Verlag, 1995.

Blanchard, Ian, Goodman, Anthony, and Newman, Jennifer, eds. *Industry and Finance in Early Modern History: Essays Presented to George Hammersley to the Occasion of his 74th Birthday*. Stuttgart: Franz Steiner Verlag, 1992.

Bloor, David. *Knowledge and Social Imagery*. Chicago: University of Chicago Press, 1991.

Braunstein, Philippe. "Innovations in Mining and Metal Production in Europe in the late Middle Ages." *Journal of European Economic History* 12:3 (1983), 573–591.

Breidbach, Olaf and Ziche, Paul, eds. *Naturwissenschaft um 1800: Wissenschaftskultur in Jena-Weimar*. Weimar: Verlag Hermann Böhlaus Nachfolger, 2001.

Brianta, Donata. "Education and Training in the Mining Industry, 1750–1860: European Models and the Italian Case." *Annals of Science* 57 (2000).

Clark, Stuart. *Thinking with Demons: The Idea of Witchcraft in Early Modern Europe*. New York: Oxford University Press, 1999.

Clark, William. *Academic Charisma and the Origins of the Research University*. Chicago: The University of Chicago Press, 2006.

——. "Der Untergang der Astrologie in der deutschen Barockzeit," in *Im Zeichen der Krise: Religiosität im Europa des 17. Jahrhunderts*, Harmut Lehmann and Anne-Charlott Trepp, eds. Göttingen: Vandenhoeck & Ruprecht, 1999.

Cohen, Claudine and Wakefield, André, eds.. *Protogaea: Gottfried Wilhelm Leibniz*. Chicago: The University of Chicago Press, 2008.

Cooper, Alix. *Inventing the Indigenous: Local Knowledge and Natural History in Early Modern Europe*. New York: Cambridge University Press, 2007.

Craddock, Paul T. *Early Metal Mining and Production*. Washington D.C.: Smithsonian Institute Press, 1995.

Cunningham Andrew, and Jardine, Nicholas, eds. *Romanticism and the Sciences*. New York: Cambridge University Press, 1990.

Darmstaedter, Ernst. *Berg-, Probir-, und Kunstbüchlein*. Munich: Verlag der Münchner Drucke, 1926.

Darnton, Robert. *Mesmerism and the End of the Enlightenment in France*. Cambridge, MA: Harvard University Press, 1968.

Dülmen, Richard von. *Kultur und Alltag in der Frühen Neuzeit: Religion, Magie, Aufklärung*. München: C.H. Beck, 1999.
Dym, Warren Alexander. "Mineral Fumes and Mining Spirits: Popular Culture in the Sarepta of Johann Mathesius (1506–1565)." *Reformation and Renaissance Review* 8.4 (2006).
——. "Alchemy and Mining: Metallogenesis and Prospecting in Early Mining Books." *Ambix* 55:3 (Nov, 2008).
Eamon, William. *Science and the Secrets of Nature: Books of Secrets in Medieval and Early Modern Culture*. Princeton: Princeton University Press, 1994.
Farr, James R. *Artisans in Europe, 1300–1914*. Cambridge: Cambridge University Press, 2000.
Foucault, Michel. *The Order of Things: An Archaeology of the Human Sciences*. London, 1974.
Fraustadt, Albert. *Geschichte des Geschlechtes von Schönberg Meissnischen Stammes*. Band I–II. Leipzig: Giesecke und Devrient, 1878.
Gierl, Martin. *Pietismus und Aufklärung*. Göttingen: Vandenhoeck und Ruprecht, 1997.
Grau, Conrad. *Die Preußische Akademie der Wissenschaften zu Berlin: Eine deutsche Gelehrtengesellschaft in drei Jahrhunderten*. Berlin: Spektrum Akademischer Verlag, 1993.
Gross, Reinhardt. *Geschichte Sachsens*. Leipzig: Edition Leipzig, 2001.
Guntau, Martin *Die Genesis der Geologie als Wissenschaft: Studie zu den kognitiven Prozessen und gesellschaftlichen Bedingungen bei der Herausbildung der Geologie als naturwissenschaftliche Disziplin an der Wende vom 18. zum 19. Jahrhundert*. Berlin: Akademie Verlag, 1984.
Hall, D.H. *History of the Earth Sciences during the Scientific and Industrial Revolutions: with Special Emphasis on the Physical Geosciences*. New York: Elsevier Scientific Publishing Company, 1976.
Hamm, E.P. "Knowledge from the Underground: Leibniz Mines the Enlightenment." *Earth Sciences History* 16:2 (1997), 77–99.
Hannaway, Owen. "Reading the Pictures: The Context of Georgius Agricola's Woodcuts." *Nuncius* 12 (1997).
Heilbron, John L. *Electricity in the 17th and 18th Centuries: A Study of Early Modern Physics*. Berkeley: University of California Press, 1979.
Heilfurth, Gerhard. *Bergbaukultur im Erzgebirge: Grundzüge und Auswirkungen*. Dresden: Sächsisches Druck- und Verlagshaus, 1995.
——. *Bergbaukultur in Südtirol*. Bozen: Verlagsanstalt Athesia, 1984.
——. *Der Bergbau und seine Kultur*. Zürich: Atlantis Verlag, 1981.
——. *Das Erzgebirgische Bergmanslied: ein Aufriss seiner literarischen Geschichte*. Frankfurt am Main: Weidlich Reprints, 1982.
——. *Bergbau und Bergmann in der deutschsprachigen Sagenüberlieferung Mitteleuropas*. Marburg: N.G. Elwert Verlag, 1967.
Heilfurth, Gerhard, and Schmidt, Leopold, eds. *Bergauüberlieferungen und Bergauprobleme im Österreich und seinem Umkreis*. Wien: Osterreichische Museum für Volkskunde, 1975.
Hermann, Armin. *Die Begründung der Elektrochemie und Entdeckung der Ultravioletten Strahlen von Johann Wilhelm Ritter*. Frankfurt am Main, 1968.
Herrmann, Walther. "Goethe und Trebra: Freundschaft und Austausch zwischen Weimar und Freiberg," in *Freiberger Forschungshefte* D9. Berlin: Akademie Verlag, 1955.
——. "Bergrat Henckel: Ein Wegbereiter der Bergakademie," in *Freiberger Forschungshefte* D37. Akademie Verlag: Berlin, 1962.
Hochadel, Oliver. *Öffentliche Wissenschaft: Elektrizität in der deutschen Aufklärung*. Göttingen, 2003.
Hoffman, Walter. *Bergakademie Freiberg: Freiberg und sein Bergbau—die Sächsische Bergakademie*. Frankfurt am Main: Wolfgang Weidlich, 1959.

Jackson, Myles W. *Spectrum of Belief: Joseph von Fraunhofer and the Craft of Precision Optics*. London: MIT Press, 2000.
Jardine, Nicholas, and Secord, James, and Spary, Emma, eds. *Cultures of Natural History*. Cambridge: Cambridge University Press, 1996.
Jobst, Wolfgang, and Schellhas, Walter. *Abraham von Schönberg—Leben und Werk: Die Wiederbelebung des erzgebirgisches Bergbaus nach dem Dreißighährigen Krieg durch Oberberghauptmann Abraham von Schönberg*. Leipzig: Deutsche Verlag für Grundstoffindustrie, 1994.
Jones, Matthew L. *The Good Life in the Scientific Revolution: Descartes, Pascal, Leibniz and the Cultivation of Virtue*. Chicago: University of Chicago Press, 2006.
Karant-Nunn, Susan C. "From Adventurers to Drones: The Saxon Silver Miners as an Early Proletariat." In *The Workplace before the Factory: Artisans and Proletarians, 1500–1800*, ed. T.M. Safley and L.N. Rosenband. Ithaca: Cornell University, 1993.
——. "Between Two Worlds: the Social Position of the Silver Miners of the Erzgebirge, c.1460–1575." *Social History* 14:3 (October, 1989), 307–322.
——. "The Women of the Saxon Mines." In *Private Lives: Women in Reformation and Counter-Reformation Europe*. Sherrin, M, ed. Bloomington, 1989.
Kasper, Hanns-Heinz, and Wächtler, Eberhard. *Geschichte der Bergstadt Freiberg*. Weimer: Hermann Böhlaus Nachfolger, 1986.
Kellenbenz, Hermann. "Technology in the Age of the Scientific Revolution 1500–1700." *The Fontana Economic History of Europe*, Carlo M. Cipolla, ed. New York: Harvester Press, 1976.
Kertz, Walter. *Geschichte der Geophysik*. New York: Georg Olms Verlag, 1999.
Knoblauch, Hubert. *Die Welt der Wünschelrutengänger und Pendler: Erkundungen einer verborgenen Wirklichkeit*. New York: Campus Verlag, 1991.
Koch, Manfred. "Paracelsus und der Bergbau." *Der Anschnitt* 16 (1964).
Kroker, Werner, and Westermann, Ekkehard., eds. *Der Anschnitt* 2, *Montanwirtschaft Mitteleuropas vom 12 bis 17 Jahrhundert: Stand, Wege und Aufgaben der Forschung*. Bochum: Deutsches Bergbau-Museum Bochum, 1984.
Labouvie, Eva. *Verbotene Künste: Volksmagie und ländlicher Aberglaube in den Dorfgemeinden des Saarraumes, 16.–19. Jahrhundert*. St. Ingbert: Röhrig Verlag, 1992.
Laube, Adolph. *Studien über den Erzgebirgischen Silbergbau von 1470 bis 1546*. Berlin: Akademie Verlag, 1974.
Laudan, Rachel. *From Mineralogy to Geology: The Foundations of a Science 1650–1830*. Chicago: The University of Chicago Press, 1987.
Lauterbach, Werner. *Bergrat Christlieb Ehregott Gellert*, in *Freiberger Forschungshefte* D200. Leipzig, 1994.
Long, Pamela. *Openness, Secrecy, Authorship: Technical Arts and the Culture of Knowledge from Antiquity to the Renaissance*. Baltimore: Johns Hopkins University Press, 2001.
——. "The Openness of Knowledge: An Ideal and Its Context in 16th- Century Writings on Mining and Metallurgy." *Technology and Culture* 32:2 (April, 1991).
Loesche, Georg. *Johannes Mathesius: Ein Lebens- und Sitten-Bild aus der Reformationszeit*. Volume I–II. Gotha, 1895.
Löscher, Hermann. "Die erzgebirgischen Knappschaften vor und nach der Reformation." *Blätter für deutsche Landesgeschichte* XCII (1956), 162–190.
Ludwig, Karl. "Georgius Agricola und Theophrast Bombast von Hohenheim genannt Paracelsus." *Der Anschnitt* 46 (1994).
Lynn, Michael R. "Divining the Enlightenment: Public Opinion and Popular Science in Old Regime France." *Isis* 92 (2001).
McClelland, Charles E. *The German Experience of Professionalization: Modern Learned Professions and their Organizations from the Early Nineteenth Century to the Hitler Era*. New York, 1991.

Meiche, Alfred. *Sagenbuch des Königreichs Sachsen*. Leipzig: G. Schönfelds Verlagsbuchhandlung, 1903.
Meinal, Christoph, ed. *Die Alchimie in der europäischen Kultur- und Wissenschaftsgeschichte*. Wiesbaden: Otto Harrassowitz, 1986.
Meixner, Heinz; Schellhas Walter; and Schmidt, Peter, eds. *Balthasar Rösler: Persönlichkeit und Wirken für den Bergbau des 17. Jahrhunderts* (Leipzig: VEB Deutscher Verlag für Grundstoffindustrie, 1980).
Multhauf, Robert P. *Neptune's Gift: A Common History of Salt*. Baltimore: Johns Hopkins University Press, 1978.
Nef, John. "Industrial Europe at the Time of the Reformation (ca 1515–ca.1540)." *The Journal of Political Economy* XLIX:1 (1941).
——. "Silver Production in Central Europe, 1450–1618." *The Journal of Political Economy*, August (1941).
Newman, William. *Promethean Ambitions: Alchemy and the Quest to Perfect Nature*. Chicago: University of Chicago Press, 2004.
——. "Technology and Alchemical Debate in the Late Middle Ages." *Isis* 80 (1989): 423–445.
Newman, William R. and Grafton, Anthony, eds. *Secrets of Nature: Astrology and Alchemy in Early Modern Europe*. Cambridge, MA: The MIT Press, 2001.
Nobis, Heribert M. "Der Ursprung der Steine: zur Beziehung zwischen Alchemie und Mineralogie im Mittelalter." In *Toward a History of Mineralogy, Petrology, and Geochemistry*. B. Fritscher and F. Henderson, eds. 1998.
Norris, John A. "The Mineral Exhalation Theory of Metallogenesis in Pre-Modern Mineral Science." *Ambix* 53:1 (March, 2006), 43–65.
Nummedal, Tara. *Alchemy and Authority in the Holy Roman Empire*. Chicago: The University of Chicago Press, 2007.
Oldroyd, David. *Thinking about the Earth: A History of Ideas in Geology*. Cambridge, Ma: Harvard University Press, 1996.
——. "Some Neo-Platonic and Stoic Influences on Mineralogy in the Sixteenth and Seventeenth Centuries." *Ambix* (1974).
Ospovat, Alexander M. "Romanticism and German Geology: Five Students of Abraham Gottlob Werner." *Eighteenth Century Life* 7 (1981–2), 105–117.
Polanyi, Michael. *Personal Knowledge: Towards a Post-Critical Philosophy*. Chicago, 1958.
Porter, Roy. *The Making of Geology: Earth Science in Britian 1660–1815*. New York: Cambridge University Press, 1977.
Porter, Theodore M. "The Promotion of Mining and the Advancement of Science: The Chemical Revolution of Mineralogy." *Annals of Science* 38 (1981): 543–570.
Priesner, Claus. *Das deutsche Salinenwesen im frühen 17. Jahrhundert*. Munich: R. Oldenbourg, 1908.
Principe, Lawrence, ed. *Chymists and Chymistry: Studies in the History of Alchemy and Early Modern Chemistry*. Sagamore Beach: Science History Publications, 2007.
Rattansi, Piyo, and Clericuzio Antonio, eds. *Alchemy and Chemistry in the 16th and 17th Centuries*. Kluwer Publishers, 1994.
Reill, Peter Hanns. *Vitalizing Nature in the Enlightenment*. Berkeley, CA: University of California Press, 2005.
Reynolds, Terry S. *Stronger Than a Hundred Men: A History of the Vertical Water Wheel*. Baltimore: The Johns Hopkins University Press, 1988.
Richter, Klaus. *Das Leben des Physikers Johann Wilhelm Ritter: ein Schicksal in der Zeit der Romantik*. Weimar: Böhlau, 2003.
Roth, Fritz. *Christian Lehmanns Leben und Werke und Seine Stellung zum Aberglauben*. Marburg: C.M. Gärtner, 1933.
Rublack, Ulinka, ed. *Gender in Early Modern German History*. New York: Cambridge University Press, 2002.

Rudwick, Martin J.S. *Bursting the Limits of Time: The Reconstruction of Geohistory in the Age of Revolution*. Chicago: The University of Chicago Press, 2005.

Shapin, Steven. *A Social History of Truth: Civility and Science in Seventeenth-Century England*. Chicago: The University of Chicago Press, 1994.

Schlechte, Horst. *Die Staatsreform in Kursachsen, 1762–1763*. Berlin: Rütten and Loening, 1958.

Schreiber, Georg. *Der Bergbau in Geschichte, Ethos und Sakralkultur*. Köln: Westdeutscher Verlag, 1962.

Schulte, Rolf. *Hexenmeister: Die Verfolgung von Männern im Rahmen der Hexenverfolgung von 1530–1730 im Alten Reich* (Frankfurt am Main: Peter Lang, 2001).

Singer, Charles. *A History of Technology*, vol 3. New York: Oxford University Press, 1954–84.

Sisco, Anneliese G., and Smith, Cyril S., trs. *Bergwerk*- and *Probierbüchlein* [orig., circa 1500 and 1520, respectively]. New York: American Institute of Mining and Metallurgical Engineers, 1949.

Smith, Pamela. *The Body of the Artisan: Art and Experience in the Scientific Revolution*. Chicago: University of Chicago Press, 2004.

Smith, Pamela H. and Schmidt, Benjamin, eds. *Making Knowledge in Early Modern Europe: Practices, Objects, and Texts, 1400–1800*. Chicago: The University of Chicago Press, 2007.

Täschner, Constantin. "Ferdinand Reich, 1799–1884: Ein Beitrag zur Freiberger Gelehrten- und Akademiegeschichte." *Mitteilungen des Freiberger Altertumvereins* 51 (1916).

Thomas, Keith. *Religion and the Decline of Magic: Studies in Popular beliefs in Sixteenth and Seventeenth Century England*. New York: Oxford University Press, 1971.

Troitzsch, Ulrich and Wohlauf, Gabriele, eds, *Technik-Geschichte: Historische Beiträge und neuere Ansätze*. Frankfurt am Main: Suhrkamp Verlag, 1980.

Tylecote, Ronald F. *The Early History of Metallurgy in Europe*. New York: Longman, 1987.

Vaccari, Ezio. "Mining and Knowledge of the Earth in Eighteenth-Century Italy." *Annals of Science* 57:2 (2000).

Veith, Heinrich. *Deutsches Bergwörterbuch*. Wiesbaden: Sändig, 1968.

Vogel, Jakob. *Ein schillerndes Kristall: Eine Wissensgeschichte des Salzes zwischen Früher Neuzeit und Moderne*. Vienna: Böhlau Verlag, 2008.

Wächtler, Eberhard and Gisela-Ruth, E., eds., *Internationales Symposium zur Geschichte des Bergbaus und Hüttenwesens, 2 Vol*. Freiberg: Bergakademie Freiberg, 1980.

Wächtler, Eberhard, ed. *Geologie und Industrielle Revolution: Studienmaterial zur Wissenschaftsgeschichte und Geschichte der Produktivkräfte*. Heft II. Freiberg: Bergakademie Freiberg, 1980.

Wagenbreth, Ottfried. *Geschichte der Geologie in Deutschland*. Stuttgart: Georg Thieme Verlag, 1999.

——. *Die Technische Universität Bergakademie Freiberg und ihre Geschichte*. Leipzig: Deutscher Verlag für Grundstoffindustrie, 1994.

Wakefield, André. *The Disordered Police State: German Cameralism as Science and Practice*. Chicago: The University of Chicago Press, 2009.

Wappler, August Friedrich. "Oberberghauptmann von Trebra und die drei ersten sächsischen Kunstmeister, Mende, Baldauf, und Brendel," in *Mitteilungen des Freiberger Altertumsvereins* 41 (1905).

——. "Alte sächsische Wünschelruten-Geschichten," in *Mitteilungen des Freiberger Altertumsvereins* 43 (1907).

Waring, George H. "The Silver Miners of the Erzgebirge and the Peasants' War of 1525 in the light of Recent Research." *The Sixteenth Century Journal* 18 (1987): 231–247.

Watanabe-O'Kelly, Helen. *Court Culture in Dresden From Renaissance to Baroque*. New York: Palgrave, 2002.
Webster, Charles. "Paracelsus and Demons: Science as a Synthesis of Popular Belief." *Scienze, Credenze, Occulte, Livelli de Cultura*. Florence, 1982.
Wegert, Karl. *Popular Culture, Crime, and Social Control in 18th-Century Württemberg*. Stuttgart: Franz Steiner Verlag, 1994.
Werner, Dietmar. *Bergmannssagen aus dem sächsischen Erzgebirge*. Leipzig: VEB Deutscher Verlag für Grundstoffindustrie, 1985.
Westermann, Ekkeharrd, ed. *Vom Bergbau- zum Industrierevier*. Stuttgart: Franz Steiner Verlag, 1995.
Wiesner, Merry E. "Guilds, Male Bonding and Women's Work in Early Modern Germany." *Gender and History* 1:2 (Summer, 1989).
Wilde, Manfred. *Die Zauberei- und Hexenprozesse in Kursachsen*. Vienna: Böhlau Verlag, 2003.
Williams, Gerhild S. *Ways of Knowing in Early Modern Germany: Johannes Praetorius as a Witness to his Time*. Burlington, VT: Ashgate Publishing, 2006.
Wilsdorf, Helmut. *Georg Agricola und seine Zeit*. Berlin: VEB Deutscher Verlag der Wissenschaften, 1956.
Winkelmann, Heinrich von. *Der Bergbau in der Kunst*. Essen: Verlag Glückauf, 1958.
Wrana, Joachim, ed. *Bergakademie Freiberg: Festschrift zu ihrer Zweihundertjahrfeier am 13. November 1965* vol. 1. Leipzig, 1965.
Ziolkowski, Theodore. *German Romanticism and its Institutions*. Princeton: Princeton University Press, 1990.

INDEX OF SUBJECTS, NAMES, AND PLACES

Studies in Central European Histories

Edited by Thomas A. Brady, Jr. and Roger Chickering

1. Blickle, P. *Communal Reformation*. The People's Quest for Salvation in the Sixteenth Century. ISBN 0 391 03730 7
2. Roper, K. *German Encounters with Modernity*. Novels of Imperial Berlin. ISBN 0 391 03695 5
3. Chickering, R. *Karl Lamprecht*. A German Academic Life (1856-1915). ISBN 0 391 03766 8
4. Brady, Jr., T.A. *Protestant Politics*. Jacob Sturm (1489-1553) and the German Reformation. ISBN 0 391 03823 0
5. Theibault, J.C. *German Villages in Crisis*. Rural Life in Hesse-Kassel and the Thirty Years War, 1580-1720. ISBN 0 391 03839 7
6. Wallace, P.G. *Communities and Conflict in Early Modern Colmar 1571-1730*. ISBN 0 391 03822 2
7. Usher Chrisman, M. *Conflicting Visions of Reform*. German Lay Propaganda Pamphlets, 1519-1530. ISBN 0 391 03944 x
8. Safley, T.M. *Charity and Economy in the Orphanages of Early Modern Augsburg*. ISBN 0 391 03983 0
9. Jackson, Jr., J.H. *Migration and Urbanization in the Ruhr Valley, 1821-1914*. ISBN 0 391 04033 2
10. Lapp, B. *Revolution from the Right*. Politics, Class and the Rise of Nazism in Saxony, 1919-1933. ISBN 0 391 04027 8
11. Scheck, R. *Alfred von Tirpitz and German Right-Wing Politics, 1914-1930*. ISBN 0 391 04043 x
12. Lovell Evans, E. *Cross and the Ballot*. Catholic Political Parties in Germany, Switzerland, Austria, Belgium and the Netherlands, 1785-1985. ISBN 0 391 04095 2
13. Anderson, A.D. *On the Verge of War*. International Relations and the Jülich-Kleve Succession Crises (1609-1614). ISBN 0 391 04092 8
14. Benz, E. *Fertility, Wealth, and Politics in Three Southwest German Villages, 1650-1900*. ISBN 0 391 04093 6
15. Bowman, W.D. *Priest and Parish in Vienna, 1780-1880*. ISBN 0 391 04094 4
16. Sun, R.C. *Before the Enemy is Within Our Walls*. Catholic Workers in Cologne, 1885-1912: A Social, Cultural and Political History. ISBN 0 391 04096 0
17. Büsch, O. *Military System and Social Life in Old Regime Prussia, 1713-1807*. The Beginnings of the Social Militarization of Prusso-German Society. Transl. by J.G. Gagliardo. ISBN 0 391 03984 9
18. Sondhaus, L. *Franz Conrad von Hötzendorf*. Architect of the Apocalypse. ISBN 0 391 04097 9
19. Patrouch, J.F. *Negotiated Settlement*. The Counter-Reformation in Upper Austria under the Habsburgs. ISBN 0 391 04099 5
20. Haude, S. *In the Shadow of "Savage Wolves"*. Anabaptist Münster and the German Reformation during the 1530s. ISBN 0 391 04100 2
21. Caldwell, P.C. & Scheuerman, W.E. *From Liberal Democracy to Fascism*. Legal and Political Thought in the Weimar Republic. ISBN 0 391 04098 7
22. Brenner, A.D. *Emil J. Gumbel*. Weimar German Pacifist and Professor. ISBN 0 391 04101 0
23. Bell, D.P. *Sacred Communities*. Jewish and Christian Identities in Fifteenth-Century Germany. ISBN 0 391 04102 9

24. Myers Feinstein, M. *State Symbols*. The Quest for Legitimacy in the Federal Republic of Germany and the German Democratic Republic, 1949-1959. ISBN 0 391 04103 7
25. Hobson, R. *Imperialism at Sea*. Naval Strategic Thought, the Ideology of Sea Power, and the Tirpitz Plan, 1875-1914. ISBN 0 391 04105 3
26. Edwards, K.A. *Families and Frontiers*. Re-creating Communities and Boundaries in the Early Modern Burgundies. ISBN 0 391 04106 1
27. Lavery, J. *Germany's Northern Challenge*. The Holy Roman Empire and the Scandinavian Struggle for the Baltic 1563-1576. ISBN 0 391 04156 8
28. Healy, R. *Jesuit Specter in Imperial Germany*. ISBN 0 391 04194 0
29. Geehr, R.S. *Aesthetics of Horror*. The Life and Thought of Richard von Kralik. ISBN 0 391 04201 7
30. Safley, T.M. (ed.). *Reformation of Charity*. The Secular and the Religious in Early Modern Poor Relief. ISBN 0 391 04211 4
31. Lindemann, M. (ed.). *Ways of Knowing*. Ten Interdisciplinary Essays. ISBN 0 391 04184 3
32. Ulbrich, C. *Shulamit and Margarete*. Power, Gender, and Religion in a Rural Society in Eighteenth-Century Europe. Transl. by T. Dunlap. ISBN 0 391 04145 2
33. Funck, M. & Chickering, R. (eds.). *Endangered Cities*. Military Power and Urban Societies in the Era of the World Wars. ISBN 0 391 04196 7
34. Beachy, R. *The Soul of Commerce*. Credit, Property, and Politics in Leipzig, 17501840. ISBN 0 391 04142 8
35. Mayes, D. *Communal Christianity*. The Life and Loss of a Peasant Vision in Early Modern Germany. ISBN 0 391 04225 4
36. Aaslestad, K. *Place and Politics*. Local Identity, Civic Culture, and German Nationalism in North Germany during the Revolutionary Era. ISBN 0 391 04228 9
37. Burnett, S.G. & Bell, D.P. (eds.). *Jews, Judaism, and the Reformation in Sixteenth-Century Germany*. ISBN 90 04 14947 3
38. Safley, T.M. *Children of the Laboring Poor*. Expectation and Experience among the Orphans of Early Modern Augsburg. ISBN 0 391 04224 6
39. Hartston, B.P. *Sensationalizing the Jewish Question*. Anti-Semitic Trials and the Press in the Early German Empire. ISBN 90 04 14654 7
40. Janik, E. *Recomposing German Music*. Politics and Musical Tradition in Cold War Berlin. ISBN 90 04 14661 X
41. Canoy, J.R. *The Discreet Charm of the Police State*. The *Landpolizei* and the Transformation of Bavaria, 1945-1965. ISBN 978 90 04 15708 8
42. Head, R.C. & Christensen, D. (eds.). *Orthodoxies and Heterodoxies in Early Modern German Culture*. Order and Creativity 1550-1750. ISBN 978 90 04 16276 1
43. Steinhoff, A.J. *The Gods of the City*. Protestantism and Religious Culture in Strasbourg, 1870-1914. ISBN 978 90 04 16405 5
44. Johnson, M.W. *Training Socialist Citizens*. Sports and the State in East Germany. ISBN 978 90 04 16957 9
45. Hanson, M.Z. *Religious Identity in an Early Reformation Community*. Augsburg, 1517 to 1555. ISBN 978 90 04 16673 8
46. Friedrich, K. & Pendzich, B.M. (eds.). *Citizenship and Identity in a Multinational Commonwealth*. Poland-Lithuania in Context, 1550-1772. ISBN 978 90 04 16983 8
47. Coy, J.P. *Strangers and Misfits*. Banishment, Social Control, and Authority in Early Modern Germany. ISBN 978 90 04 16174 0
48. Kaplan, B., Carlson, M. & Cruz, L. *Boundaries and their meanings in the history of the Netherlands*. ISBN 978 90 04 17637 9
49. Ritter, G.A. Translated by Alex Skinner. *German Refugee Historians and Friedrich Meinecke*. Letters and Documents, 1910-1977. ISBN 978 90 04 18404 6
50. Tatlock, L. (ed.). *Enduring Loss in Early Modern Germany*. Cross Disciplinary Perspectives. ISBN 978 90 04 18454 1
51. Peperkamp, E. and Rajtar, M. *Religion and the Secular in Eastern Germany, 1945 to the present*. 2010. ISBN 978 90 04 18467 1
52. Dym, W.A. *Divining Science*. Treasure Hunting and Earth Science in Early Modern Germany. 2011. ISBN 978 90 04 18642 2

www.ingramcontent.com/pod-product-compliance
Lightning Source LLC
LaVergne TN
LVHW020652300926
849161LV00004B/914